ANIMAL NUTRITION AND FEED TECHNOLOGY

PRACTICAL APPROACH

ANIMAL NUTRITION AND FEED TECHNOLOGY

PRACTICAL APPROACH

– Authors –

Dr. Vinod Kumar
Professor and Head

Dr. Muneendra Kumar
Associate Professor

Dr. Raju Kushwaha
Assistant Professor

Dr. Shalini Vaswani
Assistant Professor

Dr. Avinash Kumar
Assistant Professor

Dr. Ram Dev Yadav
Assistant Professor

Dr. Mokshata Gupta
Assistant Professor

ASTRAL
Publishing Authors Across the Globe

2025

Daya Publishing House®
A Division of
Astral International Pvt. Ltd.
New Delhi – 110 002

ISBN: 978-93-5919-479-0

Published by : **Daya Publishing House®**
A Division of
Astral International Pvt. Ltd.
– ISO 9001:2015 Certified Company –
Ground Floor, 4736/23, Ansari Road,
Darya Ganj, New Delhi-110 002
Ph. 011-43549197
e-mail: info@astralint.com
Website: www.astralint.com

Printed at : **Replika Press Pvt Ltd**

पशु चिकित्सा विज्ञान एवं पशुपालन महाविद्यालय
उत्तर प्रदेश पंडित दीनदयाल उपाध्याय पशु चिकित्सा विज्ञान विश्वविद्यालय
एवं गो अनुसंधान संस्थान, मथुरा – 281001, उ०प्र० (भारत)
College of Veterinary Science and Animal Husbandry
Uttar Pradesh Pandit Deen Dayal Upadhyaya Pashu Chikitsa Vigyan Vishwavidyalaya
Evam Go-Anusandhan Sansthan, Mathura - 281001, U.P. (INDIA)

Prof (Dr.) Vikas Pathak
M.V.Sc. PhD., FNAVSc,
FNADSc, IAAB, IIMSA
DEAN

Foreword

A Manual entitled "Animal Nutrition and Feed Technology- Practical Approach" has been prepared as per the requirements of students of B.V.Sc. & A.H. degree, M.V.Sc and Ph.D scholars. The manual will help them in understanding the preparation and processing of samples, preparation of solutions, techniques of proximate principles estimation, calcium and phosphorus determination, qualitative detection of adulterants in feeds and fodders, calculation of nutrient digestibility and ration formation for different livestock species and poultry.

I am pleased to note that the practical manual developed by the faculty members of the Department of Animal Nutrition will strengthen the academic programme and will be useful to students. I heartily congratulate all the faculty members for their sincere effort in preparing this manual.

17 January, 2025

17/1/2025
(Vikas Pathak)

Tel : 0565-2471288 (O) Cell : +91-[illegible] 9917407994
E-mail : deanvetmtr@gmail.com Website : www.upvetuniv.edu.in

PREFACE

This manual has been meticulously compiled to address the academic and practical requirements of undergraduate and postgraduate students in animal nutrition. It provides a comprehensive guide to laboratory techniques, sample preparation, chemical analysis, and feed formulation essential for livestock management and nutrition. Structured to foster both foundational understanding and advanced analytical skills, the manual introduces students to key practices such as proximate analysis, nutrient digestibility calculations, and the detection of feed adulterants. It also delves into advanced topics like ration formulation, digestion trials, and metabolic studies, ensuring a holistic grasp of feed science and technology.

Reflecting the dedicated efforts of the faculty members of the Department of Animal Nutrition, this manual draws on years of academic and research expertise. It is designed to align with curriculum standards and serve as a vital educational resource, equipping students and researchers with the analytical skills required in laboratories and feed industries. We hope this manual inspires deeper learning and innovation, serving as a valuable tool for students, educators, and researchers alike in the field of animal nutrition.

Dr. Vinod Kumar, Dr. Muneendra Kumar,
Dr. Raju Kushwaha, Dr. Shalini Vaswani,
Dr. Avinash Kumar, Dr. Ram Dev Yadav,
Dr. Mokshata Gupta

CONTENTS

CHAPTER 1

GENERAL PRECAUTIONS IN NUTRITION LABORATORY

Generally, lab safety is a matter of common sense, but there are several rules that must be followed. This chapter contains a brief synopsis of safety procedures. In addition, students must read the Standard Laboratory Procedures handout that is attached to the end of this chapter. The following material is provided as a brief summary and guide to lab safety. It does not replace assigned reading material, but gives an overview of some important points.

Table 1.1. Lab safety guide

Things	Description
Always work under instructor's or lab technician's supervision in the laboratory.	
Laboratory cleanliness	Clean and neat work areas reduces fire hazard, avoid risk of damage to clothing and books and injury from spilled chemicals.
Laboratory dress	Aprons protect clothing from corrosive or staining chemicals. Gloves protect hands from corrosive chemicals. Handle hot objects with insulated gloves. Do not wear open-toe shoes that allow spilled chemicals or broken glass to come in contact with your feet. Wear eye protection when required. Sleeves that are too tight prevent freedom of movement, whereas sleeves that are too loose may cause you to overturn apparatus or glassware.
Labelling chemicals	All chemicals should be clearly labelled. Do not use materials from unlabeled containers. Avoid contamination. Never return reagents to their container. Clearly label chemicals as you work.
Carrying chemicals and equipment	Carry long apparatus such as tubing or burettes, in an upright position close to the body. Grasp bottles firmly with both hands and hold them close to the body. Do not carry bottles by the neck. Use a bottle carrier when transporting chemicals any distance.

Transferring liquids	Remember, acid to water. Do not pipette by mouth, use a bulb. Use gloves when pouring corrosive liquids. Use a funnel when filling a bottle or flask and prevent an air block by raising the funnel. Pour hazardous liquids over a sink.
Eye protection	Normal eyeglasses are usually not adequate. Do not wear contact lenses in the lab. Eye protection is especially important when working with corrosive materials and vacuum and high pressure apparatus.
Acid/alkali spills	For acid spills, use solid sodium bicarbonate followed by water. For alkali spills, wash with water followed by dilute acetic acid.
Chemicals in the eye	Rapid treatment is vital. Run large volumes of water over eyeball until medical help is available (at least 15 minutes). Alkaline materials in the eye are extremely hazardous. Know the location of the emergency eyewash station.
Safety shower	Use this for chemical spills or a fire victim. Operate by pulling down on ring and keep the area near the shower clear at all times. Remove clothing from area affected by spills.
Extinguishing a fire	Steps in using a fire extinguisher: know its location, remove from mounting, pull pin, squeeze lever, discharge at base of flame, and use dry sand to extinguish burning metals. Rating and type of fire extinguishers used are: Type A- For ordinary combustibles; wood, paper, and cloth Type B- For flammable liquids; oil, grease, and gasoline and Type C- For use on live electrical equipment. Number on extinguisher denotes square footage the unit is capable of handling. Use a fume hood equipped with a safety glass when working with toxic or flammable materials.
Fire on clothing	Do not run or fan flames. Smother fire by wrapping victim in fire blanket or lab coat and use the shower or a carbon dioxide fire extinguisher.
Handling flammable liquids	Flammable liquids should always be stored in an approved storage cabinet. Extinguish all flames in the area where flammable solvents are used, as vapours may travel to ignition source and flash back.
Working with	
Glassware	Remove frozen glass stoppers with proper equipment. Broken or chipped glassware should be discarded. Properly support glassware with ring-stands and clamps when heating and use cork rings with round-bottom flasks.
Glass tubing	Do not touch heated glass until it has time to cool (hot glass looks just like cool glass). To remove stoppers from glass tubing or thermometers, grasp tubing close to stopper and push gently with twisting. Use water or glycerine for lubrication.

Test tube	Gently heat solids or liquids in a test tube near the liquid or solid surface. Be prepared to remove the tube from heat quickly to prevent eruption. Never point a test tube or reaction vessel at another person. For safety and neatness, place test tubes in a rack
Handling sodium and potassium	Fire or explosion may result when metallic Na or K are exposed to water. Store them under light oil. Metal can be cut safely with a spatula on a paper towel. Destroy residues with alcohol. Cool if necessary.
Protection from toxic gases	Emergency air masks should be used. However, because our lab is not equipped with such masks, clear the area where gases are, and notify the lab technician.
Waste disposal	Hot glassware or reactive chemicals should be discarded in a non-metallic container separate from paper and other flammable waste. Test-tube quantities of hazardous liquids can be flushed down the sink with plenty of water. Contact lab technician for disposal of large quantities of hazardous materials or anytime you are not sure of how something should be disposed off.
Care and handling of gas cylinder	Protect cylinder valve with cap. Fasten cylinders securely. Transport cylinders on a hand truck, don't roll. Do not drop cylinders. Mark cylinders when empty.

Note: Alerts all times: no smoking, no food or beverages, know location of exits, put books and coats in designated areas, do not leave an experiment unattended and extinguish burners when you leave the work area.

Standard procedures and safety rules

Following procedures and safety rules must be followed by personnel using the facilities of the laboratory:

1. Permission to use the facility and its equipment must be obtained before use. In the case of proposed extended use of equipment, it is recommended that such use be scheduled, in advance, with the laboratory supervisor (the use of facilities by scheduled class groups will take priority over other users).
2. Before beginning, read, entirely, all instructions from the laboratory staff regarding location and proper use of different safety equipment.
3. Personnel using the facilities of the laboratory area are required to demonstrate an understanding, and proficiency in, the use of any equipment and the conduct of any physio-chemical procedures within the premises before use, unless under direct supervision by the laboratory staff. Ask for proper instruction if in doubt about procedures.
4. Personnel using the facilities of the laboratory are required to be aware of the potential hazard involved in any procedure in which they may be engaged (fire, chemical burn, toxic fumes, electrical shock, etc.). Personnel who initiate the use of any equipment, facilities, or chemical procedures that

involve hazard, or that could become hazardous, are required to remain in that particular area until the procedure is properly terminated.

5. It is not considered good practice to work alone in the laboratory. Another person should be present or within the range of voice when any potentially hazardous procedure is being conducted.
6. It is the responsibility of all personnel using flammables to check first to ensure that the area is safe from flames and sparking equipment; it is likewise the responsibility of all personnel using flames or sparking equipment to check first to ensure that the area is free from flammables.
7. Foods and drinks are prohibited in any area of the laboratory where hazardous chemicals are in use, and eating or drinking are prohibited for all persons during whatever period of time they are engaged in usage or handling of toxic or corrosive chemicals.
8. Chemicals, equipment, and supplies are to be returned to proper storage immediately on completion of use. Desktops and work areas are to be kept free of clutter.
9. Equipment and supplies will not be removed from the premises unless properly checked out. Check out procedure for glassware, equipment, and lab space should be followed.
10. All materials, including samples, should be properly labelled. Use proper labeling tape and write clearly. Materials not properly labelled will be discarded.
11. Everyone is responsible for properly washing his or her own glassware and returning it to storage. Clear ovens, desiccators, and related equipment as soon as possible so that others may use the facilities.
12. A log book must be maintained for recording name of personnel using the facilities along with date, parameter and signature. Additionally, keep a record of supplies (chemical, glass ware, etc.) needed in the laboratory.
13. Scan bulletin boards and chalkboards in the labs daily for notices that may pertain to the use of equipment or facilities.

CHAPTER 2

FAMILIARIZATION WITH VARIOUS FEEDS AND FODDERS

To assess the nutritive value of animal feeds we should be able to identify various fodders and their use for feeding livestock. This is more important for green and dry fodder as it forms the bulk of animal feed. Apart from giving information of feeding value of feedstuff this knowledge enables to classify feedstuffs into different categories within which the substitution of feed is possible for computation of ration for livestock and poultry. In addition to fodder, various concentrates are used in the ration of animals especially milking animals and young growing stock. These concentrates comprise of cereals and their by products, oilseeds, oilcakes, leguminous grasses and roots. They differ in chemical composition which influence their feeding value, e.g., cereals are rich in starch and other carbohydrates; whereas, oilseeds and their by products are more rich in P than Ca. Therefore, to make economical use, it is essential to know the chemical composition of various feeds and fodders.

Definitions

1. **Feedstuffs:** Feed and fodder covering all edible material in livestock ration and having nutritive value are feedstuffs. These include naturally occurring plants and animal products and by products of milling and other processing and artificially prepared stuffs.
2. **Roughages:** These are bulky/low density feeds, containing more than 18% of crude fibre and less than 60% TDN. Therefore, they are low in energy content, e.g. hay, straw, hull (covering of seed), husk, stovers and other related fodders.
3. **Fodders:** The entire above ground part of nearly mature corn/sorghum/ legume plant in fresh or cured form.

4. **Forage:** Crops used in the whole plant form (except roots) like pasture, hay, green crop for feeding purpose.
5. **Energy or basal feed:** These are the grains and its by-products which contain not more than 20% crude protein and 18% crude fibre. Nutritionally, these are energy concentrates rich in starch and sugars. Average protein content of basal feeds ranges from 8 to 12% and sometimes ether extract content is less than 5%, e.g., grains of barley, maize, oat, jowar, wheat, bajra and rice.
6. **Protein supplements:** They are the products containing more than 18% of protein. Animal by-products viz. marine by-products (fish meal) and avian by-products (feather meal and poultry litter meal).
7. **Mineral supplements:** Mineral supplements are those which supply minerals like bone meal, chalk, oyster shell, shell grit, marble chip, etc.
8. **Concentrates:** They are high density feeds or feed mixtures, which are highly digestible and containing more than 60% TDN and less than 18% crude fibre.

Classification of feedstuffs

Feedstuffs are classified into three parts:

1. Roughages
2. Concentrates
3. Additives

Roughages are divided into two parts-dry and green roughages. Green roughages are sub divided into leguminous and non-leguminous. Concentrates are divided into two parts –vegetable and animal sources. Vegetable sources are sub-divided into three parts-grains (Energy Sources), oil seed cakes (Protein sources) and by products. Additives are divided into five parts- Antibiotics, anabolics, hormones, mineral supplements and vitamin supplements.

Roughages

These are bulky feeds containing more than 18% CF and less than 60% TDN. Roughages are sub-divided into two major groups-dry (>80% DM) and succulent (80% DM). The dry roughages are hays (leguminous and non-leguminous). Straw (wheat straw, paddy straw, barley straw, etc.) and stovers (bajra, jowar, maize, etc.). Except good quality leguminous and non-leguminous hays, remaining are non-maintenance type roughages. Succulent roughages include mainly cultivated leguminous and non-leguminous fodders. The most commonly used kharif leguminous fodders are cowpea and guar and non-leguminous ones are jowar, maize and sudan grass. The leguminous and non-leguminous fodders of rabi season are Lucerne, berseem, pea and oat and barley, respectively. In addition, pasture, tree leaves silages and root crops are also grouped under this head.

Concentrates

Feed or feed mixtures containing less than 18% CF and over 60% TDN are classed as concentrates. They are rich in either energy or protein and accordingly are called as energy concentrates (when CP is less than 18%) or protein concentrates (when CP value exceeds 18%). Energy concentrates are grain and seeds (maize, barley, sorghum etc.), milling byproducts (brans and chunis), roots (tapioca tubers turnip and potatoes) and molasses. Protein supplements, on the other hand may be further categorized according to the source of origin (a) animal origin protein supplements (meat meal, blood meal, fish meal, feather meal, etc.) and (b) plant origin protein supplements (oilseed cakes and meals i.e. groundnut cake, sesame cake, mustard cake, linseed oil meal etc.). Animal origin protein supplements except dried whole and skimmed milk, contain more than 47% CP, whereas those of plant origin, contain less than 47% CP except soybean oil meal (CP 49%).

In addition, a third group of feedstuffs include mineral and vitamin supplements and additives (antibiotics, hormones, etc.) which when added to the total feed mixture improve the performance of the supplemented livestock.

Classification of animal feeds

Feeding stuffs	Types	Example
1. Roughages >18% CF <60% TDN	**A) Dry** >80% DM	i) Hays (18-34% CF, 40-60% TDN & considerable carotene)
		Leguminous hay: (hays of lucerne, berseem etc.)
		>10.5% CP, >0.9% Ca
		Non leguminous hays: (oat, sorghum, doob grass etc.)
		6-10.5%, <0.9% Ca
		ii) Low quality roughages
		>28% CF, several >34% CF, <52% TDN, several <40% TDN, <6% CP, little carotene, if any.
		(straws like wheat, rice, oat, jowar, barley)
	B) Succulent <80% DM	i) Green fodder
		a) Leguminous (berseem, lucerne, cowpea, cluster bean and green pea >10.5% CP)
		b) No-leguminous (jowar, maize, bajra, oat etc. and grasses of sudan, napier, guinea, etc.)
		c) Pasture
		d) Tree leaves
		e) Silage (25-45% DM, medium in carotene)
		f) Root crops (9-30% DM, <3% CF)

2. Concentrates <18% CF >60% TDN	**A) Energy rich** (<18% CP)	a) Grain and seeds (maize, barley, sorghum etc.) b) Mill byproducts (wheat bran, rice bran, chunis of pulses) c) Roots (turnip, potatoes, tapioca etc.) d) Molasses
	B) Protein rich (>18% CP)	a) Animal origin (mostly >47% CP, >1% Ca, >1.5% P & <2.5% CF) i) Animal byproducts- blood meal, meat scraps/meal, dried whole and skimmed milk, meat cum bone meal ii) Marine byproducts-fish meal, shrimp meal iii) Avian byproducts- feather meal and manure and poultry litter b) Plant origin (mostly <47% CP, <1% Ca, <1.5% P & >2.5% CF) protein source includes defatted oil seeds (<17% EE) like cakes and meals e.g. soybean meal (SBM), ground nut cake (GNC), mustard oil cake (MOC), cotton seed cake, till cake, etc. Brewer's grain and yeast (>17% EE)
3. Mineral supplements- bone meal, dicalcium phosphate (DCP), mono calcium phosphate, phosphoric acid, chalk, oyster shell, shell grit, marble chip.		
4. Vitamin supplements- tocopherol acetate		
5. Additives- probiotics, prebiotics, antioxidants, enzymes, hormones, flavoring and coloring agents, antibiotics, sweeteners, etc.		

CHAPTER 3

PREPARATION AND PROCESSING OF SAMPLES FOR CHEMICAL ANALYSIS

In animal nutrition experiment different kinds of substances like feed, fodder, silage, urine, faeces, etc. are processed for analyzing their chemical and biochemical characteristics. This is mainly done for quality control of feedstuff for manufacture of compound feed and digestion/ metabolic trial experimentation to measure the digestibility of feed ingredients and their chemical composition.

Sampling type

1. Sampling of concentrates e.g. whole or crushed grain, mixed feeds, compound cattle feed, etc.
2. Sampling of roughages e.g. dry and green roughages.
3. Sampling of molasses.
4. Sampling of urine.
5. Sampling of faeces.
6. Sampling of milk.
7. Collection of blood samples.

Sampling of concentrates

It includes sampling of individual concentrate ingredients and processed feed like compound cattle feed, pelleted feed, feed wafers, feed crumbles, etc.

Sampling of concentrates ingredients for quality control

This is a very important step in feed industry where raw feed ingredients like grains, cakes, agro-industrial byproducts and finished product are analyzed for their chemical constituents and therefore the price of those ingredients are determined. 'Parkhi' is used for taking samples from bags. It is a tubular metallic probe having pointed end. For sampling from bulk, a deep bin probe or thermo sampler is used.

Sampling from bags

It is impossible to take sample from each of the bag containing raw material coming to feed factory. So, a minimum number is set for sampling from bag.

Number of bag in the lot	Minimum number of bags to be sampled
Upto 30 bags	All bags
31 to 300 bags	30 bags
301 to 500 bags	30 bags + 10% excess

Bags are selected at random for sampling. Samples are taken out from the bags with the help of parkhi. All the samples are pooled together and a representative sample is taken for analysis.

Sampling from bulk

Feed ingredients are generally stacked in godowns. For sampling of feed ingredients from bulk sampler is inserted upto a desired depth into different places of bulk called spots. Number of spots depends upon the quantity of ingredient stored.

Bulk quantity	Number of spots for sampling
Upto 30 tonnes	30 spots
31 to 100 tonnes	50 spots
101 to 300 tonnes	50 spots+ 10% extra

After that all the spot samples are mixed and a representative sample is prepared. Representative sample is again divided into three sets. Each set contains three polythene bags containing at least 150 gm of representative sample. One set is kept with the buyer, second set is going to the seller and third one is kept at a mutually agreed place. This last set used if some dispute occurs between two parties.

Sampling of dry roughages

Roughage may be of different type and size i.e. long and chaffed. In case of long fodder like hay (except legume hay), grass, un-chaffed straw or stover, the sample can be drawn by hand from different parts of stacks. For small stack of about 100 quintals there should 1 meter distance between two spots. If the stack is of 10 to 100 tonnes, the distance can be of 4 to 5 meters. For chaffed dry fodder like wheat straw the sampling can be done by a large inverted scoop with a long handle so that the scoop can be inserted upto a certain depth.

The primary samples of long fodder are chaffed. Similarly primary chaffed sample are mixed thoroughly. Three sets of secondary sample are packed into three bags of 1.5 kg each and then distributed as in previous case.

Sampling of feeds, fodders and other biological materials during digestion or metabolic trial

For precise estimation of chemical composition of feed stuff given to animal, sampling of each feed component is required. During digestion/metabolic trial, chemical composition of both offered and residual feed stuff, faeces and urine should be estimated. Thus sampling of the above substances is necessary. In case of lactating animals, milk samples are also collected.

Sampling of green fodder

Samples are taken from the green fodder offered should be collected just before feeding to the animals. The fodder is chaffed thoroughly and about 200-300 g sample is taken in polythene packet. The packet should be closed immediately and DM content of the sample is measured as early as possible to avoid any error due to evaporation.

Sampling of dry roughage

After mixing the chaffed dry roughage thoroughly 100 gm sample is taken in a polythene sheet. 50 gm sample is taken for dry matter analysis and rest 50 gm is kept in numbered polythene bag. Such 6 to 8 days (digestion trial) samples are collected kept for further proximate analysis.

Sampling for silage/haylage

It is cut into small pieces of 3-5 cm with hand chopper. After that it is mixed thoroughly and 500 gm of sample is taken in a polythene bag. The bag is closed immediately to avoid moisture loss. Fresh sample is used for dry matter, ammonia nitrogen and volatile fatty acid analysis. Sample is kept for further analysis.

Sampling of concentrate

Concentrate sample is taken after mixing the whole sample thoroughly. 100 gm of sample is kept in a polythene bag and closed properly.

Sampling of feed residue

Feed residue is the left over feed by the animal after a day's (24 hours) feeding. The left over is mixed thoroughly and about 200 gm of sample is taken in a polythene bag and closed immediately.

Sampling of faeces

Total faecal matter is collected for 24 hours in a plastic or metal container. The collection should be done just after voiding of faeces. This is mixed thoroughly and 500 gm of mixed sample is taken in a polythene bag. In some species of animal like camel, sheep, goat, rabbit and deer, the faeces is in the form of pellets. The pellets should be broken down properly before mixing the whole faeces. Dry matter content of the sample is estimated as early as possible to avoid any error due to evaporation. Dried sample is kept for further analysis. For estimating nitrogen content of faeces, a particular amount (depends on total faeces voided) of faeces is collected in a wide mouth container and 5 ml of 25% of H_2SO_4 is added.

Sampling of urine

Whole urine is collected in a big plastic bottle of 5 to 10 litre capacity. 25 ml of diluted H_2SO_4 (1:4) should be kept in that container before starting the collection to prevent the loss of volatile nitrogen content of urine. Total urine excreted for 24 hours is mixed thoroughly and about 200 ml sample is taken in a plastic bottle and closed properly.

Sampling of milk

50 ml of milk sample from each animals and kept in a plastic bottle. 2 to 3 drops of potassium dichromate or mercuric chloride is added into it as a preservative and the bottle is kept in deep freeze for further analysis.

Sampling of water

Sampling of water sample is required during metabolic trial. A fixed amount of (50-100 gm) drinking water sample is kept in a polythene bottle or glass bottle with stopper.

Processing and preservation of samples

It takes some time for analysis of sample after collection. So it is required to preserve and process those samples.

Processing of dried sample:

100 gm of representative sample is dried at hot air oven at 100± 1°C and grinded in laboratory mill. The ground sample is kept in polythene bag and closed. A level is pasted over the bag detain about sample number, kind of study and date of sampling. For gross energy estimation sample is dried at 60±1°C.

Processing of wet sample:

Wet faeces sample is mixed with a known volume H_2SO_4 (25% v/v) and kept in a wide both bottle with proper closing.

Processing of urine sample:

A know amount of aliquot is stored in a Kjeldahl flask after proper numbering for nitrogen estimation. 30 ml concentrated H_2SO_4 is added immediately. Sample should be kept in duplicate. For gross energy estimation, 5 ml of urine sample is collected in stoppered glass or polythene tube and 2-3 drops of mercuric chloride is added as preservative. The tubes are stored in refrigerator.

Processing of milk sample:

Fat in milk is estimated by Gerber's method on the day of collection. For other estimation milk is preserved adding 2 to 3 drops of potassium dichromate or mercuric chloride and kept in deep freeze.

CHAPTER 4

PREPARATION OF SOLUTIONS

Solutions of accurately known strength are called standard solutions. A standard solution contains a known weight of reagent in a definite volume of solution. Molecular weight and atomic weight of commonly used chemicals has been shown in table.

Molar solution

Molar solution is one, which contains one molecular weight of the reagent in one litre of the solution. Molarity is expressed as M.

Normal solution

Normal solution is one, which contains one equivalent weight of the reagent in one litre of the solution. Normality is expressed as N. Equivalent weight of acid = Molecular weight/ No. of replaceable H ions.

Table 4.1. Description of commonly used chemicals

Name	Formula	Mol. wt. (g/mol)	Eq. wt. (g/mol)
Hydrochloric acid	HCl	36.5	36.5
Nitric acid	HNO_3	63	63
Potassium hydroxide	KOH	56	56
Calcium hydroxide	$Ca(OH)_2$	74	37
Potassium dichromate	$K_2Cr_2O_7$	294	49
Sodium thiosulphate	$Na_2S_2O_3.5H_2O$	248	248
Sodium chloride	$NaCl$	58.5	58.5
Potassium chloride	KCl	74.5	74.5
Iodine	I_2	254	127

Most of the synthetic dyes in generally used as indicators are organic substances of complex structure. Among the most reliable of these indicators are methyl red and phenolphthalein.

Table 4.2. Common indicators used in animal nutrition laboratory

Indicator	pH range	End point	Preparation
Methyl red	4.4 to 6.3	Pink in acidic medium and colourless in basic medium	For preparing a stock solution, 0.2 g of dye is dissolved in 100 ml of alcohol and filtered
Methyl orange	2.9 to 4.0	Orange in acidic medium & colourless in basic medium	For preparing a stock solution, 0.1 g of dye is dissolved in 100 ml of distilled water, filtered and used
Phenolphthalein	8.3 to 10.0	Pink in basic medium and colourless in acidic medium	For preparing a stock solution, 0.2 g of phenolphthalein is dissolved in 110 ml of alcohol and 90 ml of distilled water

Preparation of standard solutions commonly used for analysis in animal nutrition

1. Preparation of N/10 H_2SO_4

Equivalent weight of H_2SO_4 = 49 g, specific gravity = 1.84 g/ml

So, volume of 49 g H_2SO_4 = 26.6 ml

Concentrated H_2SO_4 (reagent grade) is about 97 per cent pure.

Therefore, actual amount of concentrated H_2SO_4 required for 1.0 litre of N/10 H_2SO_4

$$1\ N\ H_2\ SO_4 = \frac{100}{97} \times 26.6 = 27.42\ ml$$

Thus, for 1.0 litre of N/10 H_2SO_4 solution, 2.74 ml of concentrated H_2SO_4 is required.

Procedure

Take 2.74 ml sulphuric acid in a beaker half-filled with distilled water. Transfer the contents and washings to a volumetric flask (1litre) and make volume up to the mark. Shake well and titrate this solution with 10 ml of 0.1 N Na_2CO_3 using mixed / methyl orange as an indicator. Repeat the titration to get at least three concordant readings.

Standardization

Suppose 10 ml of 0.1 N Na_2CO_3 = 9.5 ml of H_2SO_4

$V_1 N_1 = V_2 N_2$

$10 \times 0.1 N = 9.5 \times N = 0.10526$

To prepare 1 litre N/10 H_2SO_4, the volume of 0.10526 N acid required is 1000 x 0.1/0.10526 = 950 ml. Take 950 ml of 0.10526 N acid and dilute it to one litre. Check it again with N/10 Na_2CO_3 for three times. It must neutralize equal volume of N/10 Na_2CO_3 solution. Label it as 0.1 N H_2SO_4.

Precautions

Add H_2SO_4 with the help of a burette. Never add water to an acid.

2. Preparation of N/10 NaOH solution

Molecular weight of NaOH is 40, acidity (number of replaceable OH group) is 1 therefore; equivalent weight of NaOH is 40. Thus, 4 g of NaOH dissolved in one litre of solution will give N/10 NaOH solution.

Procedure

Weigh quickly 4 g NaOH in a beaker (as it is hygroscopic) and dissolve it in distilled water (preferably CO_2 -free). Transfer the contents and the washings to a volumetric flask (1 litre). Cool and then make volume up to the mark. Shake well and standardize this solution against N/10 oxalic acid using phenolphthalein as an indicator. Label it as 0.1 N NaOH solution.

3. Preparation of N/10 $KMnO_4$ solution

Dissolve 3.2 g $KMnO_4$ in one litre of distilled water. Boil it for 10-15 minutes and then allow to stand for few days and then filter it through glass wool. Take 10 ml of N/10 oxalic acid in a beaker. Add 5 ml dilute sulphuric acid, warm it to 60-70°C and titrate against $KMnO_4$ from the burette till a light pinkish colour appears. Take three concordant readings.

Suppose, 10 ml 0.1 N oxalic acid = 9.75 ml of $KMnO_4$

$V_1 N_1 = V_2 N_2$

$10 \times 0.1 N = 9.75 \times N_2$

$$N_2 = \frac{10 \times 0.1 \text{ N}}{9.75} = 0.10256$$

To prepare 1000 ml 0.1 N $KMnO_4$ solution, the volume of $KMnO_4$ will be taken.

$$= \frac{100 \times 9.75 \times 0.1}{10 \times 0.1}$$

Now, take 975 ml of prepared $KMnO_4$ solution and make it 1000 ml by adding distilled water.

Note: Ordinary or even pure distilled water contains traces of organic matter which reduces the $KMnO_4$ solutions. That is why the solution is boiled and kept for some time before standardization. In the absence of sufficient amount of dilute

H_2SO_4 or due to the rapid addition of $KMnO_4$ in titration flask, brown turbidity (manganous oxide) may appear.

4. Preparation of N/10 Na_2CO_3 solution

Molecular weight of Na_2CO_3 is 106, Na_2CO_3 = 2HCl = 2NaCl+H_2O+CO_2

So, acidity of Na_2CO_3 is 2 and equivalent weight of Na_2CO_3 is 53. Therefore, 5.3 g Na_2CO_3 is required for each litre of solution to make N/10 Na_2CO_3.Na_2CO_3 is hygroscopic; therefore, it must be made perfectly anhydrous before it is weighed out. Quantity of acid/ alkali required for preparation of different molar/normal solutions has been shown in Table 4.3.

Table 4.3. Acid or alkali required for preparation of different normal solutions

Normality	95% NaOH (g/l), M=N	36% HCl (ml/l), M=N	98% H_2SO_4 (ml/l), M=2N	<68% HNO_3 (ml/l), M=N
0.1 N	4.2	8.5	2.7	6.5
0.5 N	21.0	42.6	13.7	32.8
1.0 N	42.1	85.3	27.4	65.7
1.5 N	63.1	127.9	41.1	98.5
2.0 N	84.2	170.6	54.8	131.4
2.5 N	105.2	213.2	68.5	164.2
5.0 N	210.5	426.5	137	328.5
10.0 N	421	853	274	657

M = Molar; N = Normal

Procedure

Take 6-7 g of Na_2CO_3 (A.R.) in a nickel crucible and heat it in a hot air oven at about 100°C for few hours so as to drive out any moisture and to convert any moisture and to convert any preformed $NaHCO_3$ to Na_2CO_3. Cool in a desiccator and weigh exactly 5.3 g dried salt and dissolve it in a little quantity of freshly boiled distilled water. Transfer it to one litre measuring flask and make volume up to the mark. Shake well and label it as 0.1 N Na_2CO_3 solution.

5. Preparation of N/10 oxalic acid

$COOH_2$. $2H_2O$ (Oxalic acid) is to be dissolved in one litre of distilled water to get N/10 oxalic acid solution.

Procedure

Weigh accurately 6.3 g $(COOH)_2$. $2H_2O$ and transfer it to a volumetric flask (1 litre), half-filled with distilled water. Shake well and make the volume up to the mark. Label it as N/10 oxalic acid solution.

Note: If anhydrous oxalic acid (COOH) is available then dissolve 4.5 g of the acid in one litre of distilled water to get 0.1 N oxalic acid solution.

6. Preparation of standardized 0.313 N (1.25%) NaOH solution

Add 13.16 g of NaOH (95% NaOH) in one litre distilled water and shake well. Standardize this solution against known concentration of oxalic acid solution using phenolphthalein as an indicator.

7. Preparation of standardized 0.256 N (1.25per cent (w/v) H_2SO_4 solution

To prepare 1.25 per cent (w/v) H_2SO_4 solution, 12.5 g of H_2SO_4 (100 per cent) is to be added to distilled water to make the volume 1000 ml. Volume of H_2SO_4 to be taken:

$$-\frac{12.5 \times 100}{1.84 \times 97} = 7 \text{ ml}$$

Procedure

Add 7.0 ml concentrated H_2SO_4 (specific gravity 1.84 g/ml and 97 per cent concentration) in a 1000 cc volumetric flask half-filled with distilled water. Shake well and add distilled water to make volume up to the mark. Standardize this solution against known concentration of Na_2CO_3 using mixed/methyl orange indicator.

Precaution and preservation of standard solutions

The bottle must be kept tightly stoppered to prevent evaporation of solvent. Some solutions must be protected from atmospheric gases. For example, sodium hydroxide solution is affected by atmospheric CO_2.

$2NaOH + CO_2 = Na_2CO_3 + H_2O$

But never put a glass stopper on NaOH solution container because NaOH will react with air and glass between stopper and neck of volumetric flask. It will fix permanently and you cannot remove the glass stopper from volumetric flask. $KMnO_4$ solution should be preserved in colour (amber) bottles. The container should be shaken well before the withdrawal of a portion of solution to ensure uniform composition of the solution.

CHAPTER 5

WEENDE OR PROXIMATE SYSTEM OF FEEDS AND FODDERS ANALYSIS

A method for the quantitative analysis of the different nutrients in feed is the Weende or proximate analysis, based on the Weende analysis that was developed in 1865 by Wilhelm Henneberg and Friedrich Karl Adolf Stohmann at Weende experimental station in Germany. Under the scheme various nutrients that have some common properties are grouped together and analyzed. Proximate Analysis is a partitioning of compounds in a feed into six categories based on the chemical properties of the compounds. The six categories are:

1. Moisture
2. Ash
3. Crude protein (CP)
4. Ether extract (EE; crude fat)
5. Crude fibre (CF)
6. Nitrogen-free extracts (NFE)

1. Moisture

It gives information about the characteristics of the feed, whether it is succulent or dry feed stuff. It also gives the information about suitability of storage of feedstuff. The feed stuffs containing more than 15% of moisture are not safe to be stored in bulk as there will be growth of fungus and mould. Fermentation may also take place which lead to combustion. Determination of water content in feed is important since the relative cost of feed per unit of nutritional value is also calculated on dry matter basis. For preservation of fodder as silage, determination of water is essential since best silage can only be prepared when the water content in the fodder is between 60 to 65%.

2. Ash

It is the inorganic residue which is estimated after the sample is burnt at 600°C. The part of ash which is soluble contains all mineral elements and high content of insoluble ash indicates poor quality of feed because it represents sand and silica. Total ash content is negatively correlated with energy content of feed. Higher values of acid insoluble ash indicate adulteration of feed with sand and silica.

3. Crude protein (CP)

For estimation of crude protein in the feeds and fodder, nitrogen content is estimated. It is multiplied with 6.25 factor, as protein contains about 16% of nitrogen in its molecule. Determination of crude protein in the feedstuff is important since they are classified according to protein content, e.g., oil cakes. legumes and their byproducts are rich in protein content. Similarly, products from animal sources like fish meal, meat meal, etc. are high in crude protein content. Fodders like wheat straw, paddy straw, jowar straw, etc. are poor in crude protein content. Crude protein also gives indirect information about feed quality, the digestible energy of the feed. Where the protein content is high, the crude fibre content is usually low and therefore digestibility of the fodder will be higher.

4. Ether extract (EE)

Ether extract is that fraction of feedstuff which is obtained when the feed stuff is subjected to continuous extraction with petroleum ether. This includes, besides the fats, cholesterol, plant sterols, lecithin, chlorophyll and substances which are not true fats and are known as pseudo-fats. The fat in the feeds and fodder is important as concentrated source of energy. It also supplies some of the essential fatty acids like oleic, linoleic, linolenic and arachidonic acids and fat soluble vitamins to animals. The ether extract is variable in various feeds and fodders. Green forages are rich in ether extract than their by-products like straws. Expeller processed oil cakes contain 3-5% ether extract, ghani processed oil cakes contain 7-8% ether extract whereas solvent extracted cakes contain very little amount of ether extract (0.5-1%). The ether extract is determined by using Soxhlet's ether extraction apparatus.

5. Crude fibre (CF)

It is that fraction of total carbohydrates which is not digested after successive boiling with dilute acid and dilute alkali. Crude fibre consists of cellulose, hemicellulose, pentosans and small fractions of lignin. The bulkiness of feed is correlated with the crude fibre. Bulk is important for the animal for satisfaction of hunger. Crude fibre absorbs moisture in the intestine and gives normal distention to digestive tract. This promotes the peristaltic movements of intestine which is helpful for voiding the digesta from the intestine. Roughages contain high proportion of crude fibre and are indigestible part of ruminant ration. Concentrates contain less proportion of crude fibre. Crude fibre content of feed is negatively correlated with its nutritive value.

6. Nitrogen free extracts (NFE)

It is that fraction of total carbohydrates which is obtained when sum of moisture, CP, EE, CF and ash is subtracted from 100. It includes monosaccharide, disaccharides, trisaccharides and some of soluble polysaccharides. Part of lignin also comes in this fraction as lignin is soluble in alkali.

Main features of Weende's system of analysis:

1. It is approximating feed value without use of animal.
2. Lignin is indigestible and gets added into NFE as it is partly soluble in NaOH.
3. It is recognized by AOAC as method of feed analysis and accepted as a tool in quality evaluation of feed by BIS.
4. Carbohydrate is divided into two fractions: highly digestible NFE and less digestible crude fibre.
5. This scheme (crude fibre estimation) represents the process of digestion in alimentary tract of animals.
6. The highly digestible fraction calculated by difference is supposed to represent soluble carbohydrate.
7. It is a very simple system of analysis accepted worldwide.

The basis of Weende's analysis system is that feedstuff contains organic and inorganic constituents. The organic matter consists of carbohydrates (CF and NFE), fats, proteins and other nitrogenous compounds. The carbohydrate is divided into two groups, i.e., the CF which is less digestible and NFE which is easily digestible. NFE includes complete polysaccharides along with some soluble lignin. Hence, in this system NFE is over estimated.

Advantages of proximate analysis

This method divides feeds into broad categories such as roughages with high CF and concentrates with low CF. This method establishes the basis for classifying livestock feed for chemical analysis. The chemical composition value of Indian feedstuffs given by ICAR is based on this system of evaluation of feedstuff. The proximate analysis helps in quality control and detecting adulteration of feed ingredients. This method is used as too for purchase of feed ingredients and is very simple, economic and rapid.

Limitations of proximate analysis

1. This method results in underestimation of CF, since part of lignin and hemi cellulose gets dissolved during alkali digestion, which in turn leads to overestimation of NFE.
2. NFE is calculated by difference; hence any experimental error may change the NFE value.

3. The proximate principles are not chemically nutritionally pure but are the mixture of substances having common properties; hence it is termed as proximate principles.
4. In estimation of ether extract, along with true fats (triglycerides), the pseudo-fats also get estimated.
5. In protein estimation, along with true proteins, the NPN substances also get estimated; hence it is called crude protein.
6. Minimum importance is given to micro-nutrients.
7. The practical aspects like palatability and cost are not considered.

CHAPTER 6

ESTIMATION OF DRY MATTER (DM) CONTENT

Principle

When feed samples are subjected to the temperature of 100 ± 5° C in hot air oven till the constant weight is attained, all the moisture gets evaporated as vapour and what remains behind is known as dry matter.

Apparatus Required

1. Analytical balance
2. Hot air oven
3. Desiccator
4. Moisture cup /Aluminum tray
5. Pair of tongs

Procedure

1. Chop or grind the sample received in the laboratory to particular size. Weigh the moisture cup/tray accurately. Take 5 to 10 g for air-dried samples like cereals and for green 50-100 g of green forage in cup/tray. Weigh it. The temperature of hot air oven for moisture estimation is 100 ± 5° C.
2. In some feeds having high decomposition matter like sugars, higher temperature is not desired; hence, the range of temperature should be 70 to 80° C.
3. Many a times, special arrangements are done to estimate the moisture content of the samples like silages containing volatile fatty acids, otherwise there might be over estimation of moisture.
4. For the samples which are liquid or pasty in nature, e.g., molasses, estimation of moisture is done by modified method.

5. The samples which are classified as tubers have to be chopped down in thin slices and without any loss of time it should be transferred into hot air oven.
6. Wilting of the sample takes place if not transferred immediately. Chopping is done for exposing more surface area. We should not lose time in between the sample received in the laboratory and transferring it to the hot air oven, otherwise it may lead to reduction in moisture percent by process of wilting. This is true in case of samples like greens and silages.
7. Time required for drying ranges from 6 to 24 hrs according to the nature of sample. It is very necessary to observe the temperature range of 100±5 °C. Since the excess of temperature will remove not only moisture but also destroy other nutrients.

Actual steps followed in estimation of moisture of a sample in the laboratory

8. Weigh the empty tin or tray.
9. Take the sample in tray and weigh it again.
10. Transfer this weighed tin with sample to hot air oven and adjust the temperature of hot air oven to 100 ±5 °C. Keep it tree for 12 to 24 hrs.
11. After heating for particular time take out the tin with sample cool it and weigh it. The correct criterion is heating it till the constant weight is obtained. Loss in weight of tin with sample is due to loss of moisture from sample.

Observations

1. Weight of empty tray = W g.
2. Weight of tray + sample before drying = W_1 g.
3. Weight of sample taken = W_1- W g.
4. Weight of tray + sample after drying = W_2 g
5. Loss in weight = W_1-W_2 g.

Calculation

$$\text{Moisture \%} = \frac{\textit{Loss in weight}}{\textit{Weight of sample}} \times 100$$

$$\text{Moisture \%} = \frac{W_1 - W_2}{W_1 - W} \times 100$$

DM % = 100 - Moisture percent in the feed

Precaution:

- Weighing should be quick and accurate.

CHAPTER 7

ESTIMATION OF TOTAL ASH CONTENT

Principle

Ash is the inorganic or mineral component of the sample left after complete ignition of the sample at 600±30° C in muffle furnace. It can be partitioned into soluble portion comprising of mineral matter and insoluble ash consisting of silica or sand. It can be used for preparing mineral extract for the estimation of various minerals and finding out OM and NFE by difference.

Apparatus Required

1. Analytical balance
2. Desiccator
3. Silica basin
4. Tongs
5. Muffle furnace

Procedure

1. Dry the crucible in an oven. Cool it in desiccator and record its weight.
2. Weigh by difference about 5-10 g of oven dried sample in this crucible.
3. Decarbonize the sample on a burner or on an electrical heater till the smoke ends up. Precaution is taken so that during charring the feed sample does not catch fire, and feed material in silica basin is not disturbed by any means.
4. After complete charring, transfer the crucible to a muffle furnace with the help of a pair of long tongs. Keep it at 600° C for two hours (At this temperature all organic matter will be burnt leaving white ash).
5. Remove the crucible from the furnace, cool it in a desiccator and weigh.

The previously recorded empty crucible weight is now subtracted and the weight of ash is, thus, determined.

6. Preserve this ash for extraction of the acid soluble minerals and, thus, for estimation of the acid insoluble ash.

Observations

1. Weight of empty silica basin = W_1 g
2. Weight of silica basin + oven dried sample = W_2 g
3. Weight of sample on DM basis = W_2-W_1 g
4. Weight of silica basin + Ash = W_3 g
5. Weight of total ash = W_3 - W_1 g

Calculations

$$\text{Total Ash \%} = \frac{\text{Weight of total ash}}{\text{Weight of sample on DM basis}} \times 100$$

$$= \frac{W_3 - W_1}{W_2 - W_1} \times 100$$

Precautions

- The ash is highly hygroscopic and thus weighing should be done quickly.
- Switch off the fans and put on the exhaust while decarbonizing the sample.
- If more samples are ashed at a time, note the pattern of silica basin placed inside the muffle, as there is probability of loss of identification during ashing.

CHAPTER 8

ESTIMATION OF ACID INSOLUBLE ASH (AIA) CONTENT

Principle

When ash is digested with dilute HCl (1:1), inorganic salts are dissolve in it. That portion of the total ash which is insoluble in dilute HCl is called acid insoluble ash. It contains sand and silica, which has no physiological role to play in animal body. It is useful to know the extent of contamination while handling feedstuffs as well as it help in detecting degree of adulteration. Plant materials like straw, husk, stem and stalks are highly lignified or woody and are rich in silica. Sand may also find entry into feeds due to faulty handling in the field or used as an adulterant.

Equipments and glasswares required

1. Hot air oven
2. Desiccator
3. Silica basin
4. Tongs
5. Muffle furnace
6. Volumetric flask
7. Funnel
8. Beaker (250 ml)

Reagents

1: 1 HCl

Procedure

1. Prepare the total ash.
2. To the silica basin containing total ash add 20-40 ml of 1:1 HCl. Cover the crucible with watch glass and digest for 20 to 30 minutes on water bath.
3. Remove the watch glass and rinse with water in to the contents of the crucible. After the digestion, the color of content of the silica basin will change to yellow.
4. Cool the silica basin. Take Whatman filter paper 42, fold it twice and put it on funnel. Filter the contents of silica basin through the filter paper in volumetric flask. Wash the silica basin several times with hot distilled water till it becomes acid free.
5. Test the filtrate drops at the tip of the funnel with blue litmus paper. After making it acid free make the volume to 250 ml in a volumetric flask by adding cold distilled water.
6. Label it with necessary details. The same aliquot is used for further analysis. Take out the filter paper and transfer it to silica basin and keep it in the hot air oven at 135° C for 1 hour.
7. After complete drying char the filter paper and transfer the silica basin in muffle furnace at 600±30° C for 2-3 hrs. After complete ashing remove the silica basin, put it in desiccator. Cool it to room temperature and weigh it.

Observations

1. Weight of empty silica basin = W_1 g
2. Weight of silica basin + dried sample = W_2 g
3. Weight of sample = W_2-W_1 g
4. Weight of silica basin + insoluble ash = W_3 g
5. Insoluble ash = $W_3 - W_1$ g

Calculations

$$\text{AIA (\%)} = \frac{\text{Weight of insoluble ash}}{\text{Weight of dried sample}} \times 100$$

$$= \frac{W_3 - W_1}{W_2 - W_1} \times 100$$

Precaution

- Do not make volume in the flask while its contents are hot.

CHAPTER 9

DETERMINATION OF CRUDE PROTEIN (CP) CONTENT

CP content of feeds, fodders and faecal samples can be determined by using manual method i.e. micro Kjeldahl method or semi automatic method

1. Micro Kjeldahl method

Principle

The protein content is estimated on the assumption that all the nitrogen in the feed exists in the form of protein. As the protein contains 16% nitrogen, the value for crude protein is obtained by multiplying the nitrogen content by the protein factor (100/16 = 6.25). The nitrogen compounds in the sample are converted to ammonium sulphate by acid digestion with concentrated boiling H_2SO_4 acid. The acid digest is cooled and diluted with distilled water and steam distilled with a strong alkali (sodium hydroxide). The released ammonia is condensed into a boric acid mixed indicator. The ammonia in the boric acid is quantified by titration with standard H_2SO_4 acid.

Digestion

Organic nitrogen+ conc. $H_2SO_4 = (NH_4)_2SO_4$

Distillation

$(NH_4)_2SO_4 + 2\ NaOH = 2\ NH_3 + Na_2SO_4 + 2\ H_2O$

$2\ NH_3 + H_2SO_4 = (NH_4)_2SO_4$

Equipments

1. Kjeldahl digestion apparatus
2. Kjeldahl flasks
3. Distillation assembly

4. Erlenmeyer flask (250 ml capacity)
5. Burette

Reagents

1. Concentrated H_2SO_4 acid (reagent grade)
2. 40% sodium hydroxide (40% NaOH)
3. Digestion mixture [copper sulphate (act as catalyst): sodium sulphate or potassium sulphate (raise boiling point) 1: 9]
4. Boric acid mixed indicator or Tashiro Indicator (20 ml methyl red and 10 ml bromocresol green in 1000 ml of 2% boric acid solution).
5. N/7 H_2SO_4 acid solution.

Procedure

1. Digestion

a. Take 1.0-2.0 g of representative sample and transfer it to Kjeldahl flask.
b. Add 20-30 ml conc. H_2SO_4.
c. Add 2-5 g digestion mixture.
d. Place the flask on digestion bench and heat till the solution becomes clear blue.
e. Remove the flask and cool it.

2. Distillation

a. Allow flask to cool and then gently add about 5-10 ml distilled water to the Kjeldahl flask and transfer the solution to 100 ml volumetric flask with repeated washings till the volume is reached.
b. Take 10 ml aliquot and transfer it to micro-Kjeldahl assembly.
c. Add just sufficient volume (15-20 ml) of 40% NaOH solution to the distillation assembly to make the contents alkaline. Put stopper immediately.
d. Allow the distillation for 10-15 minutes.
e. Steam distills the mixture and collect 125 ml of distillate in a 25 ml boric acid mixed indicator taken in an Erlenmeyer flask.

3. Titration

a. Remove the conical flask after washing tip of the condenser with distilled water into the flask.
b. Titrate contents of the Erlenmeyer flask with N/7 H_2SO_4 solution.

Calculation

$$CP\ (\%) = \frac{\text{ml of N/7 } H_2SO_4 \text{consumed } \times 0.002\ \times 6.25}{\text{Weight of dried sample}} \times 100$$

Note: 0.002 is nitrogen equivalent of standard H_2SO_4 acid i.e. 1 ml of Standard N/7 (0.143 N H_2SO_4 acid = 0.002 g or 2 mg)

The equivalent is derived like this: $2NH_3 + H_2SO_4 = (NH_4)_2SO_4$.

In the equation, 2 units of nitrogen is reacting with 1 unit of H_2SO_4= (14 x 2) 28 units of nitrogen is reacting with 98 units of H_2SO_4. Therefore, 7 units of H_2SO_4can react with 2 units of nitrogen or 7 mg (contained in 1 ml of N/7 H_2SO_4 acid) of H_2SO_4 acid is equivalent to 2 mg of nitrogen.

Semi automatic method (KEL-PLUS)

The advance semi automatic nitrogen /protein analysis system is a pollution free and environmental friendly system. The hazardous acid fumes liberated during digestion are efficiently dissolved in water through giant aspiration filter pump and sent to the drain without polluting the atmosphere. The compact design of automatic block digestion with built in exhaust fume manifold system avoids the use of large fume hoods or cupboards, saving installation cost as well as space. The instrumentation includes automatic microprocessor based temperature controller with various built in safety features to enable a clean and a safe operation.

The second stage of experiment feeds the digested sample in an automatic distillation system. The system can handle hot digestion tubes directly without the need of transferring the sample. This avoids manual spillage error during transfer, which would otherwise happen in a conventional traditional system. The Distillation process is fully programmable with precise automation and the entire process of sample dilution, reagent addition, steam injection, Boric acid addition, residue removal, water refilling, goes unhindered with a single start operation and it takes not more than 3 to 6 min. for the completion of distillation.

The third stage of the equipment a titration phase is simple and easy to handle with conventional glass burette titration system by any user.

Principle

The crude protein can be determined by pelican's KEL-PLUS system, which is an improved semi-automated method of analysis.

Glass wares

Block digestion system (Kelplus KES 06L) - heating block, insert rack, manifold rack, support rack for insert rack, support rack for manifold rack, exhaust manifold, filter pump. Distillation apparatus (Kelplus-Elite ex), scrubber unit (Kelplus-kelvac).

Procedure

Weigh the sample in a digital precision balance and transfer it into the digestion tube. Add digestion mixture to the sample. Also add required volume of concentrated H_2SO_4 into the tube Using SS insert rack, take samples to the digestions systems Manifold are permanently connected to Kelvac. Place the manifold with manifold rack over the digestion tube. The water inlet to Kelflow is connected to water tap. Load the insert rack with tubes and manifold into the block digestion system. Open the water tap for switch on the Kelvac system. Ensure proper sealing of exhaust manifold to avoid leakage of fumes Switch on the power switch. Set the temperature in the temperature controller. If frothing is observed in the sample, digest the sample in low temperature for some time and increase the set temperature.

After the digestion of samples is over, separate the sample from the system. Remove the insert rack with samples and place it over the tube support rack. Allow water to flow through the Kelflow for some time, to remove the fumes from the tube. Remove the manifold from the tubes. Place the manifold in the manifold support rack. Take out the sample one by one and dilute with distilled water. Dilution is necessary to avoid solidification of sample on cooling.

The diluted sample is placed in the distillation system for ammonia recovery. Keep the alkali, reservoirs to the rear of the system. Immerse the hose coming out from the system into the alkali tank wait for the READY indication in the control panel. Open the water inlet to the water condenser. Select the volume and alkali to be added using selector switch and press the alkali switch for alkali addition. Select the distillation time by using selector switch and press process switch for steam addition. Ensure addition of reagents and steam. Reagents can also be added manually. Water refilling to the boiler takes place automatically. During refilling of water ensure an alarm and an indication. Ensure the distilled water reservoir always filled with sufficient distilled water. Ensure collection of ammonia in the receiver solution.

Remove the receiver solution with ammonia and take to a auto titrator / manual burette for titration Titrate either by direct or back titration to estimate the nitrogen percentage.

Calculation

$$\text{Nitrogen (\%)} = \frac{14 \times \text{Normality of acid (TV} - \text{BV)}}{\text{Weight of dried sample} \times 1000} \times 100$$

Where: TV is the titration value, BV is the blank value

% Protein = % Nitrogen x 6.25

Reactions

Digestion: Organic matter + $H_2SO_4 \rightarrow CO_2 + H_2O + (NH_4)_2SO_4 + SO_2$

Distillation: $(NH_4)_2SO_4 + 2\ NaOH \rightarrow 2\ NH_3 + 2\ H_2O + Na_2SO_4$

$NH_3 + H_3BO_3$ (boric acid) $\rightarrow NH^{4+} + H_2BO_3^-$ (borate ion)

Titration: $H_2BO_3^- + H^+ \rightarrow H_3BO_3$

Common problems

Like any other method, the Kjeldahl procedure is subject to problems with methodology, technique and equipment. Understanding the principles behind the Kjeldahl process and using appropriate quality control procedures should prevent most problems from occurring.

Common problems include the following: Sample size and type, incompatibility with amounts and types of acid, salt or catalyst used samples, standards and reagents. Inadequate or excessive distillation, dilution volume or aliquot volume; foaming or bumping during digestion, uneven digestion, Inadequate or excessive sample temperature, precipitation of salt or caking of digested sample, leaks in apparatus connections, inadequate NaOH addition, improper standardization instrumental and calculation errors, etc.

CHAPTER 10

ESTIMATION OF ETHER EXTRACT (EE) CONTENT

Estimation of EE content using Soxhlet apparatus

Principle

The EE is that fraction of feedstuff which is obtained when the feed stuff is subjected to continuous extraction with petroleum ether. This includes besides the fats, cholesterol, free fatty acids, lecithin, chlorophyll, resins, volatile oils and plant sterols. These substances other than crude fats are called pseudo-fats. Since the proportion of the components of the EE will be variable in composition in different foods, the energy supply from the ether extract in the oil-cakes would be mostly the fats but in the green fodders large portion of chlorophyll, volatile oils, carotenoids, etc. would also be there. The EE content is variable in various feeds and fodders.

The Ether Extract is determined by Soxhlet's ether extraction apparatus. The ether extract content of the sample is estimated by extracting it with a fat solvent like petroleum ether, benzene, chloroform, diethyl ether, etc. Ether is continuously volatilized at 55-60^{0} C, condensed and allowed to pass through the sample in a Soxhlet apparatus.

Apparatus required

a. Soxhlet's extraction assembly: This consists of three parts:

1. Condenser at the top
2. The Soxhlet's or extractor siphon in the middle
3. The receiver flask at the bottom.

   These three parts are assembled by means of ground glass interchangeable joints

b. Extraction thimble

c. Balance

d. Hot plate with temperature controlling device.

Chemical required

Petroleum ether (boiling point- 40-60° C)

Procedure

1. Weigh about 2-5 g moisture free sample and transfer it in a thimble.
2. Plug the mouth of the thimble with fat free absorbent cotton.
3. Take the clean dry receiver flask from the Soxhlet's assembly and weigh it accurately. Introduce this thimble in the Soxhlet's extractor. The height of the thimble should be such that highest point of it should remain below the upper point of the siphoning tube.
4. Fill the extractor 1.5 times of its capacity with petroleum ether by pouring through the glass funnel and assemble the Soxhlet's apparatus.
5. Plug the tip of the condenser with cotton. Place the apparatus on a hot plate at 40° C. Fix it by clamps to a retort stand and start cold water circulation in the condenser.
6. Extract approximately for 8 hours (about 250 cycles).
7. After extraction is over, remove the thimble with the material from the extractor. Assemble the apparatus again and heat it on the hot plate to recover all the ether from the receiver flask.
8. The flask now contains only the crude fat and traces of petroleum ether. Disconnect the receiver flask, wipe the outside of the flask thoroughly with a clean dry cloth to remove the film of moisture and dust and dry it in a hot air oven at 100° C, for 1 hour, cool it in a desiccator and weigh.

Observations

1. Weight of empty thimble/filter paper = W g
2. Weight of thimble/ filter paper + dried sample = W_1 g
3. Weight of dried sample taken = W_1 - W g
4. Weight of empty receiver flask = W_2 g
5. Weight of receiver flask + ether extract = W_3 g
6. Weight of ether extract = W_3 - W_2 g

Calculation

$$\text{EE (\%)} = \frac{W_3 - W_2}{W_1 - W} \times 100$$

Precautions

- While placing the thimble containing sample in the Soxhlet flask, be sure that the top of the thimble is above the siphon tube.
- Put a cotton swab on the mouth of the condenser to avoid the loss of either vapours.

Estimation of EE content using Soxtec system

Introduction

The Soxtec system has been developed to make solvent extraction faster, safe and more rational. The design of the Soxtec system makes it possible to extract crude fat from feeds in typically 45 minutes, to process 6 samples simultaneously and to use only 20 ml of solvent per extraction. The heating system with no electricity close to the extraction unit has minimized possible explosion risks.

Equipment

1. Analytical balance
2. Soxtec system
3. Thimbles 26 x 60 mm
4. Heating oven
5. Desiccator

Chemicals

1. Appropriate solvent e.g. diethyl ether, petroleum ether, hexane.
2. Cotton wool (defatted)
3. Boiling chips (glass ball – 4-5 mm)

Soxtec HT procedure

1. Grind the samples thoroughly.
2. Load each thimble with about 3 g (W_1) of the well mixed sample and cover with a thin layer of cotton wool.
3. Dry the thimbles.
4. Insert the thimbles into the Soxtec HT.
5. Dry and pre-weigh (W_2) the extraction cups (with boiling chips). Add 25-50 ml of the solvent into each cup.
6. Insert the cups in to the Soxtec HT.
7. Extract for 15 minutes in "Boiling" position and for 30-45 minutes in "Rinsing" position.
8. Evaporate the solvent.

9. Release the cups and dry at 100°C for 30 min.
10. Cool the cups in a desiccator and weigh (W_3). Calculate percentage fat/oil according to the formula.
11. Fat extraction can also be done by using Soxtherm system

$$EE\ (\%) = \frac{(W_3 - W_2)}{W_1} \times 100$$

CHAPTER 11

ESTIMATION OF CRUDE FIBRE (CF) CONTENT

Determination of crude fibre – Muslin cloth method

Principle

When moisture and fat free sample is boiled first with dilute acid (1.25% H_2SO_4) and then with dilute alkali (1.25% NaOH) the soluble carbohydrates and protein go into solution and are extracted. The undissolved residue left behind represents cellulose; hemicellulose and lignin which put together are called as crude fibre. When the residue after drying and weighing is heated at 600° C, the fibre content is obtained by deducting the weight of ash from the weight of dried residue.

Apparatus required

1. Spoutless beaker 600 ml capacity.
2. Condensing unit.
3. Vitrosil basin.
4. Heating plate
5. Weighing balance.
6. Horn spatula
7. Muslin cloth.
8. Erlenmeyer's flask.
9. Glass funnel.
10. Dessicator.
11. Measuring cylinder.
12. Wash bottle.

13. Beaker
14. Pair of tongs.
15. Tripod stand.
16. Clay pipe triangle.
17. Hot air oven.

Chemicals Required

1.25% H_2SO_4 and 1.25% NaOH

Procedure

1. Weigh accurately about 2-3 g of moisture and fat free sample on a previously weighed glazed paper.
2. Transfer the sample in tall spoutless beaker with a mark made at 200 cc level with a glass pencil.
3. Add 200 ml of 1.25 % H_2SO_4 in the beaker containing sample and immediately connect a flask with reflux condenser and heat so that the contents begin to boil. Continue the boiling for exactly 30 min. During boiling keep the volume constant. To avoid foaming add few drops of iso-amyl alcohol, if necessary.
4. Filter the content of a beaker in a conical flask through muslin cloth (18 threads/sq cm) held over the glass funnel. Wash the residue on cloth with boiling water till residue gets acid free. Test with blue litmus paper.
5. Transfer the material from muslin cloth back to the original beaker. Add 1.25 % NaOH 200 cc up to the mark and heat to boil for exactly 30 min.
6. Filter the contents of beaker through the same muslin cloth on the conical flask. Wash the residue with boiling water till residue gets alkali free Test with red litmus paper. Dip the muslin cloth with residue in petroleum ether for 5 minutes to remove the fat.
7. Transfer the residue to a clean dry silica basin Dry it at 100 ± 5° C in hot air oven for 2-3 hours, cool and weigh.
8. Decarbonize and ignite the residue in muffle furnace at 600° C for 2-3 hrs. Cool and weigh. The loss of weight due to ignition is the weight of crude fiber.

Observations

1. Weight of glaze paper = W_1 g
2. Weight of glaze paper + dried sample = W_2 g
3. Weight of dried sample = W_2-W_1 g
4. Weight of silica basin + digested dried residue = W_3 g
5. Weight of silica basin + ashed residue = W_4 g
6. Loss in weight = W_3- W_4 g

Calculation

$$CF\ (\%) = \frac{\text{Loss in weight}}{\text{Weight of dried sample}} \times 100$$

$$= \frac{(W_3 - W_4)}{W_2 - W_1} \times 100$$

Determination of crude fibre using Fibertec system

Reagents

1. Sulphuric acid – 0.256 N (12.5 g of H_2SO_4 diluted to 1 litre and mixed)
2. Potassium hydryoxide – 0.223 N (12.5 g of KOH dissolved in de-ionized water and diluted to 1 litre)
3. n-Octanol
4. Acetone

Procedure

1. Weigh and transfer quantitatively an exact quantity (about 1 g = Wo) of fat free sample into clean filter crucibles placed in a crucible stand.
2. When a set of crucibles is loaded with samples, hook the stand on the front of the hot extraction unit, fix the crucible holder to the crucible and transfer them to the boiling positions in the hot extraction unit.
3. Through the valves add to each sample 150 ml of 0.256 N sulphuric acid, preheated in one of the reagent heating systems.
4. Add a few drops of octanol to prevent foaming and heat to boiling.
5. Adjust heat and boil for 30 minutes.
6. Filter and, if necessary, remove sediment from the filter surface by applying reversed pressure.
7. Wash three times with hot de-ionized water from the spray device sliding on the bar above the boiling tubes. Use about 30 ml of water each time and suck as dry as possible.
8. Add to each sample 150 ml of potassium hydroxide solution preheated in the second reagent system.
9. Add a few drops of octanol and boil as above for another period of 30 minutes.
10. Filter and wash as above three times with hot water.
11. Using the crucible holder move the crucibles from the hot extraction unit to the cold extraction unit.
12. Wash three times with acetone (about 25 ml each time) and suck dry.

13. Remove the crucibles and transfer them to a crucible stand.
14. Dry the crucibles (in the stand) at 100 °C overnight or at 130 °C for 2 hours.
15. Cool the crucibles in desiccator or weigh them hot directly from the drying oven. Where the latter procedure is used, check zero of balance after each weighing and adjust weight accordingly (Weight = W_1).
16. Ash the sample in the crucibles at 500 °C for at least 3 hours.
17. Cool the crucibles slowly to 100 °C or room temperature depending on which weighing method you use, and weigh again (weigh = W_2).
18. Fibretherm can also be used for estimation of fibre content.
19. Rinse the crucibles by blowing the ash out with compressed air or if the crucibles are dirty, wash with chromic acid.

Calculation

$$CF\ (\%) = \frac{(W_1 - W_2)}{W_0} \times 100$$

CHAPTER 12

CALCULATION OF NITROGEN FREE EXTRACT (NFE) CONTENT

Nitrogen-free extract (NFE) consisting of carbohydrates, sugars, starches, and a major portion of materials classed as hemicellulose in feeds. This is determined by subtracting the percentage of CP, EE, CF and total ash on DM basis and including moisture on as such basis from 100, respectively.

Calculation

NFE (% as such basis) = 100 – (CP%+ EE%+ CF%+ Total Ash%+ Moisture %)

NFE (% on DM basis) = 100 – (CP%+ EE%+ CF%+ Total Ash% on DM basis)

Note

1. Since NFE is not analyzed directly and determined by difference, cumulative error occurred during analysis of other ingredients may reflect in NFE content.
2. Organic matter can also be calculated by addition of percentages of CP, CF, EE and NFE.
3. Total carbohydrate content of feed can be determined by addition of percentages of CF and NFE.

CHAPTER 13

DEMONSTRATION OF DETERGENT METHODS OF FORAGE ANALYSIS

Though proximate analysis system is the starting point of chemical analysis, yet it has certain limitations. The major limitations lie with crude fiber (CF) and nitrogen- free extract (NFE). As per proximate analysis, CF should contain total cellulose, hemicelluloses and lignin. However, only part of the hemicellulose and lignin comes out in the crude fiber fraction. The remaining portion shows as NFE, which is usually thought consisting mainly of highly digestible sugars and starch. Consequently, the hemicellulose, low in digestibility and lignin essentially indigestible appears in NFE fraction, thereby giving an incorrect index regarding nutritive value of a given feed.

To overcome these limitations, P.J. Van Soest (1963 and 1967) developed a new scheme of analysis of forage dry matter in terms of cell wall and cell constitutions using various detergents. This method divides the dry matter of plant cell into a) cell wall content and b) cell contents. According to this scheme, cell contents, which are soluble in neutral detergents, include lipids, sugars, organic acids, starch, soluble proteins, non protein, nitrogenous compounds and other water soluble matters. While the insoluble material known as neutral detergent fiber (NDF) comprises cell wall constituents which include hemicellulose, cellulose, lignin, lignified nitrogenous compound, heat damaged proteins, keratins and silica. When NDF is treated with acid detergent, hemicellulose, some cell wall nitrogenous compounds and acid soluble ash pass into solution leaving all other cell wall constituents in the insoluble fraction which is named as acid detergent fibre (ADF) or lignocellulose. This lignocellulose when treated with 72% H_2SO_4, the cellulose is dissolved in the acid leaving behind the lignin and acid insoluble ash, from which lignin can be determined by loss of ignition. Detailed for the estimation of these fraction are given below:

Determination of neutral detergent fibre (NDF) content

Principle

It is rapid method for estimation of the cell wall constituent in vegetable feed stuff. It divides the dry matter of feed ingredients into completely soluble and insoluble materials. The insoluble material in neutral detergent solution is cell wall contents of feed. The cell contents are determined from the differences between percent cell wall and 100.

Apparatus required

1. Refluxing apparatus: It contains one spout-less beaker of 500 ml capacity and a round bottom flux as a condenser.
2. Sintered glass crucibles of 50 ml capacity with course porosity (Grade I)
3. Weighing balance
4. Vacuum pump
5. Hot plate
6. Wash bottle
7. Hot air oven and muffle furnace

Reagents required

1) Neutral detergent solution (NDS)

a. Distilled water – 1litre
b. Sodium lauryl sulphate – 30 g
c. Disodium ethylene diamino tetra acetate (EDTA) dehydrate – 18.61 g
d. Sodium borate decahydrate – 6.81 g
e. Disodium hydrogen phosphate – 4.56 g
f. 2- ethoxy ethanol – 10 ml

Make two separate solutions. One will contain EDTA and sodium borate decahydrate and other will contain sodium lauryl sulphate and 2- ethoxy ethanol. Mix those solutions. Put disodium hydrogen phosphate in a beaker and add some distilled water in it. Heat it until it gets dissolved. Then add it to the previous solution containing other ingredients.

2) Decahydronaphthalin (Decalin) – reagent grade

3) Acetone

Procedure

1. Take 0.5 to 1.0 g air dry and ground sample into a beaker of refluxing apparatus.
2. Add 100 ml NDS in the beaker. Reflux the content for 60 minutes from the onset of boiling.

3. Filter the content in sintered glass crucible by using vacuum pump.
4. Wash the beaker and filtrate thrice with hot distilled water. Wash twice with acetone.
5. Dry the crucible at 100° C for overnight and weigh it (W1).
6. Ignite the residue in muffle furnace at 600° C for 2-3 hrs. Cool and weigh (W2).
7. Calculate the NDF content as percent cell wall constituents and estimate cell soluble material by subtracting the value from 100.

Observation

Weight of dried sample = A g

Wt. of crucible +dry sample = W1 g

Weight of crucible + ash = W2 g

Calculation

$$\text{NDF (\%)} = \frac{W1 - W2}{A} \times 100$$

Cell contents (%) = 100 – cell wall constituents (NDF %)

Determination of acid detergent fibre (ADF) content

It is a method to determine the ligno-cellulose content in feed stuffs. The different between NDF and ADF is an estimate of hemi-cellulose content of feed. It is used as a preparatory step to estimate the lignin content in animal feed stuffs.

Apparatus required

1. Refluxing apparatus: It contains one spout-less beaker of 500 ml capacity and a round bottom flux as a condenser.
2. Sintered glass crucibles of 50 ml capacity with course porosity (Grade I)
3. Weighing balance
4. Vacuum pump
5. Hot plate
6. Wash bottle
7. Hot air oven and muffle furnace

Reagents required

a. 1 N H_2SO_4

b. Cetyl trimethyl ammonium bromide (CTAB) solution [20 g CTAB is added to 1000 ml 1 N H_2SO_4 and stirred well, the solution is called acid detergent solution (ADS)].

c. Decalin (reagent grade)
d. Acetone

Procedure

1. Take 0.5 to 1.0 g air dry and ground sample into a beaker of refluxing apparatus.
2. Add 100 ml acid detergent solution and 2 ml decalin in the beaker. Heat to boiling for 5 to 10 minutes. Reduce the heat and reflux the content for 60 minutes from the onset of boiling.
3. Filter on previously weighed sintered glass crucible by using vacuum pump.
4. Wash the beaker and filtrate thrice with hot distilled water. Wash twice with acetone.
5. Dry the crucible at 100^0 C for overnight and weigh it (W1).
6. Ignite the residue in muffle furnace at 600° C for 2-3 hrs. Cool and weigh (W2).

Observation

Weight of dried sample	= A g
Wt. of crucible +dry sample	= W1 g
Weight of crucible + ash	= W2 g

Calculation

$$\text{ADF (\%)} = \frac{W1 - W2}{A} \times 100$$

Determination of acid detergent lignin (ADL) content

Principle

The ADF residue consists of cellulose, lignin, cutin and acid insoluble ash. Treatment of ADF residue with 72 percent sulphuric acid dissolves cellulose. Ashing of the residue will determine the crude lignin fraction including cutin.

Apparatus required

1. Refluxing apparatus: It contains one spout-less beaker of 500 ml capacity and a round bottom flux as a condenser.
2. Sintered glass crucibles of 50 ml capacity with course porosity (Grade I)
3. Weighing balance
4. Vacuum pump
5. Hot plate
6. Wash bottle

7. Hot air oven and muffle furnace
8. Glass tray

Reagents required

72% H_2SO_4 by weight.

Procedure

1. Prepare the acid detergent fibre.
2. Place the crucible on the glass tray. Keep one end of the 2 cm higher so that the acid can be drained away from the crucibles.
3. The crucible content is covered with cooled (15 °C) 72% H_2SO_4 and stirred with a glass rod to a smooth paste. Fill the crucible almost half with acid and stir. Let glass rod remain in the crucible. Refill the crucible with 72% H_2SO_4 and stir at hourly interval as the acid drained away. Crucible does not need to be kept full at all times. Three additions are sufficient.
4. Keep crucible overnight at 100 °C and take weight.
5. Ignite the crucible in muffle furnace at 500 to 550 °C for 3 hours. Cool it and take the weight of crucible and ash.

Observation

Wt. of dry sample	= A g
Weight of crucible + residue left after acid dilution	= B g
Weight of crucible + ash	= C g

Calculation

$$\text{ADL (\%)} = \frac{B - C}{A} \times 100$$

CHAPTER 14

ESTIMATION OF CALCIUM (Ca) CONTENT

Principle

When a solution containing Ca is treated with saturated ammonium oxalate ($C_2H_8N_2O_4$ or $NH_4OOCCOONH_4$), all the Ca present is precipitated as calcium oxalate (CaC_2O_4). The precipitate, when treated with conc. H_2SO_4 gets dissolves forming calcium sulphate ($CaSO_4$) and liberating free oxalic acid ($C_2H_2O_4$) which is quantitatively estimated by titrating against N/10 $KMnO_4$ solution to arrive at the Ca content present in the given sample.

1 ml of N/10 KMnO4 = 0.002 g of Ca

Apparatus required

1. Beaker 250 ml
2. Heater or Bunsen burner
3. Conical flasks
4. Filtration stand
5. Glass rod
6. Burette stand
7. Test tubes
8. Wash bottle
9. Graduate pipettes (10 ml, 20 ml)
10. Whatman filter paper 40
11. Glass funnel
12. Test tube rack

Chemicals required

1. Diluted HCl
2. Diluted sulphuric acid
3. 5% ammonium hydroxide
4. Methyl red indicator (Dissolve 0.1 g of methyl red in 100 ml of 95% alcohol)
5. Saturated ammonium oxalate
6. Calcium chloride solution
7. 0.1 N $KMnO_4$ solution

Procedure

1. Pipette out an aliquot 10 ml from the stock ash solution into 250 ml beaker.
2. To this add 10 ml saturated solution of ammonium oxalate and 2-3 drops of methyl red indicator.
3. Mix solution thoroughly with glass rod. The solution turns red as the ash has been extracted with HCl in the preparation of the acid soluble ash.
4. The solution which is acidic in nature is neutralized by adding dilute 5 % ammonium hydroxide solution drop by drop to just reach the pH range in the alkaline medium (till color becomes faint pink).
5. Cover the beaker and keep on hot plate at 100°C. Heat till vapors start coming. This makes the precipitate of calcium oxalate granular.
6. Allow standing at least for 4 h or preferably overnight to settle the precipitate.
7. Filter through Whatman filter paper No. 40 and wash the precipitate with hot water till it is free from soluble oxalate*.
8. Give repeated washings to filter paper with distilled water.
9. Transfer filter paper back to original beaker & give washing with distilled water so that calcium crystals come down.
10. Give wash with 10 mL of 1:9 H_2SO_4 on filter paper followed by distilled water
11. Heat the beaker till the vapours start coming.
12. Titrate the solution (when it is hot, not go below 80° C) against N/10 $KMnO_4$ solution till stable pink color appears.
13. Add filter paper to same beaker and crush. Titrate again and record the volume of N/10 $KMnO_4$ used for titration.

***{Test of soluble oxalates: Take 5-10 ml $CaCl_2$ solution in a test tube and add few drops of filtrate into it. Formation of precipitate indicates that it is not free from soluble oxalates}.**

Calculation

$$\text{Ca}\left(\%\right) = \frac{\text{ml of } \frac{N}{10} \text{ KMnO}_4 \text{required} \times 0.002 \times \text{AF}}{\text{Weight of dried sample}} \times 100$$

$$\text{AF (aliquot factor)} = \frac{\text{Volume of aliquot prepared}}{\text{Volume of aliquot taken}}$$

Chemical reactions

The Calcium in 'soluble ash' is precipitated as calcium oxalate. This is then dissolved in dilute sulphuric acid.

$Ca(COO)_2 + H_2SO_4 \rightarrow CaSO_4 + (COOH)_2$ (oxalic acid) (oxalic acid is a slow reducing agent)

The Ca is determined volumetrically by titrating the solution with standard $KMnO_4$.

$2KMnO_4 + 3H_2SO_4 + 5\,(COOH)_2 \rightarrow 2MnSO_4 + K_2SO_4 + 10CO_2 + 8H_2O$

The reaction involved in titration with $KMnO_4$ is

$5Ca(COO)_2 + 2KMnO_4 + 8H_2SO_4 \rightarrow 5CaSO_4 + 2MnSO_4 + K_2SO_4 + 10CO_2 + 8H_2O$

Ca is most abundant in animal body. Approximately 99 % of Ca of the body is distributed in bones and teeth. It is important constituent of skeleton and teeth. In the bone the ratio of Ca to P is 2:1. Bone ash consists entirely of Ca and P salt. In order to assess the Ca and P nutrition of the animals the bone ash is taken as one of the measures. Ca is essential constituent of most living cells and tissue fluids.

CHAPTER 15

ESTIMATION OF PHOSPHORUS (P) CONTENT

Principle

P is estimated by volumetric method. The P is converted to "phospho-ammonium-molybdate" by addition of ammonium molybdate. The precipitate is then dissolved in excess of standard alkali. The quantity of alkali required to dissolve the precipitate is indirectly found out by back titrating with standard HCI or HNO_3. Major portion of P in the animal body is distributed in the bones. In addition to this it is also preset in soft tissues and body fluids. The P is a component of various enzyme systems and being a component of AMP, ADP and ATP it plays a very important role in carbohydrates metabolism.

Apparatus required

1. Beaker: 250 ml
2. Heater or Bunsen burner
3. Conical flasks
4. Filtration stand
5. Glass rod
6. Burette stand
7. Glass tubes
8. Wash bottle
9. Graduated pipettes (10 ml and 20 ml)
10. Whatman filter paper no. 42
11. Glass funnel
12. Test tube rack

Chemicals Required

1. 20% ammonium molybdate
2. 3% KNO_3
3. 0.1N NaOH
4. 0.1 N HNO_3
5. Phenolphthalein indicator
6. Blue litmus
7. Conc. HNO_3

Procedure

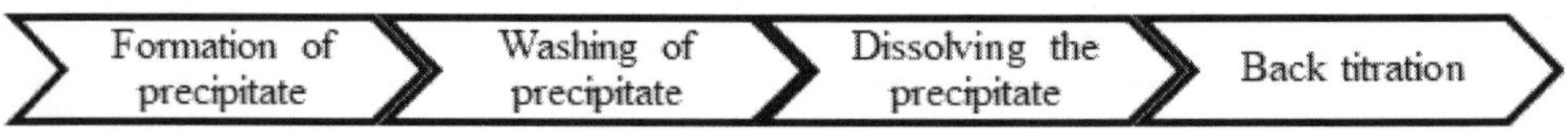

1. Take 10 ml aliquot from mineral extract (acid soluble ash; see exercise 8) in 250 ml beaker.
2. Make volume 100 ml by adding distilled water.
3. Take 10 ml of 20% ammonium molybdate (precipitant) in a test tube.
4. Take 10 ml of conc. HNO_3 in another test tube.
5. Pour both these reagents in a common stream in the beaker containing aliquot.
6. Stir the contents of beaker vigorously with glass rod till canary yellow colour precipitate is formed.
7. Allow the beaker to stand for overnight.
8. Transfer the supernatant fluid to the filter paper (Whatman filter paper no. 42) with the help of conical flask and funnel. As the filtration proceeds, continue transferring the supernatant liquid in small portions.
9. When all the supernatant liquid is thus transferred to the funnel, wait for complete filtration. The next step is washing of the precipitate.

Reasons for washing the precipitate

Ammonium molybdate is made by mixing concentrated HNO_3 with molybdic acid and ammonium hydroxide. When the reagent is added, the excess HNO_3 will be physically bound to the precipitate. Since the estimation involves a reaction between the precipitate and standard HNO_3. The excess acid has to be removed by washing.

Washing of the precipitate

3% KNO_3 is used to wash the acid from the precipitate. KNO_3 is amphoteric in nature (i.e. it behaves both as acid and alkali). Use 10 ml pipette to transfer 10 ml of 3% KNO_3 on the filter paper containing the precipitate. Mix well. Allow to settle.

Transfer the supernatant fluid to the filtering funnel. Wait for complete filtration. By this time transfer another 10 ml of 3% KNO_3 to the beaker, mix well and allow it to settle. When the previous filtration is complete, transfer the supernatant to the filtering funnel (don't start filtration unless the previous filtration is complete). This is essential for quick washing and accurate results. Continue the washing and filtration till the filtrate is acid free. Test the filtrate by taking blue litmus if litmus paper turns red. Precipitate is not acid free and further washing is needed.

Dissolving the Precipitate

When the filtrate is free from acid, gently take the filter paper and make it into a pulp. Now wash the sides of the beaker with distilled water. The entire content are now yellow. Add known amount of 0.1N NaOH till the yellow colour disappears and the contents become colourless except for the white filter paper residues. Stir well and add a drop or 2 of phenolphthalein indicator. Pink colour appears indicating excess of alkali.

Back Titration

The amount of excess alkali added is found out by back titration with 0.1N HCI or HNO_3. From the total amount of 0.1N NaOH added, the quantity of acid required for titration is deducted to arrive at the actual quantity of alkali required for dissolving the precipitate. The end point of titration is pink to colourless. Note the initial and final readings.

Observations

1. Weight of dried feed sample taken for ashing = W g
2. Volume of mineral extract (soluble ash) prepared =A
3. Volume of aliquot taken for estimation = B
4. Volume of N/10 NaOH taken for dissolving ppt = X ml
5. Volume of N/10 HNO_3 or HCL required in back titration = Y ml
6. Volume of N/10 NaOH required for dissolving ppt = X-Y ml

Calculations

1 ml of 0.1N NaOH = 0.0001347 g of P

$$P\,(\%) = \frac{0.0001347 \times \text{ml of } \frac{N}{10} \text{ NaOH required} \times AF}{\text{Weight of dried sample}} \times 100$$

$$AF\ (\text{aliquot factor}) = \frac{\text{Volume of aliquot prepared}}{\text{Volume of aliquot taken}}$$

CHAPTER 16

QUALITATIVE DETECTION OF UNDESIRABLE CONSTITUENTS AND COMMON ADULTERANTS OF FEEDSTUFFS

Adulteration is defined as the international admixture of a pure substance with some cheaper and low quality substances to make money. Costly feed ingredients like oil cakes and feed products from animal origin like fish meal are adulterated.

In animal feeds, the most common adulterants are cardboard powder, husks and hulls to increase the weight of the material. In oilseed cakes castor husks or poor quality cheap cakes may be added. Poor quality cakes may be sprayed with urea to increase their CP content. In fishmeal excessive salts may be sprinkled while drying it, to increase the total weight. Hence there is a need to detect the various adulterants present in the feed/feedstuffs. In addition naturally occurring anti-nutritional factors are also encountered in feedstuffs namely oxalates, tannins, saponins and prussic acids, etc.

Table 16.1. Common adulterants of different feed ingredients

Feed ingredient	Adulterant
Groundnut cake	Groundnut husk, urea, non- edible oil cakes
Mustard cake	*Argimona maxicana* seeds, fibrous feed ingredients
Soy bean meal	Urea
Deoiled rice bran	Ground rice hulls, saw dust

Feed ingredient	Adulterant
Fish meal	Common salt, urea
Mineral mixture	Common salt, marble powder, sand, lime stone
Molasses	Water

Detection of adulterants

Salt in fish meal

The feed ingredients, which are sold in ground form e.g. maize gluten meal, rice polish, fish meal, etc., are normally adulterated with common salt to increase total weight of the ingredient. Very high amount of common salt (NaCl), present especially in jawala fish, can cause problem when added in poultry rations. Salt in fish meal (more than 4%) cause many abnormalities in the birds e.g. diarrhoea, salt poisoning, etc.

Principle

Chloride ions, present as potassium chloride, are detected by treating the water extract of feed sample with standard silver nitrate. The intensity of brick red precipitate formed indicates the presence of salt in the sample.

Materials required

1/10 N silver nitrate, saturated potassium chromate, heater, beaker, conical flask, pipette, glass rod, iron stand, burette and muslin cloth.

Procedure

Transfer about 2 g ground sample in a beaker and add 90 ml hot water. Stir it with a glass rod. Keep it aside for 5 to 10 minutes. Filter the extract through muslin cloth. Take 10 ml filtrate in a beaker and add 3 to 4 drops saturated potassium chromate solution as an indicator. Add drop wise 0.1 N silver nitrate with a dropper in to the beaker. Formation of brick red colour precipitate indicates the presence of chloride in the sample. Higher the concentration of brick red colour precipitate, more the salt in the sample.

Alternate test

Reagent: 4% silver nitrate solution in water

Procedure

- Take 2-3 g of feed sample in a beaker.
- Add about 40 ml water and stir the contents to dissolve common salt, if present.
- Filter the contents through Whatman filter paper no 1.

- Take 2 to 3 ml of filtrate in a test tube.
- In another test tube take 2 to 3 ml of distilled water.
- Add about 3 ml of $AgNO_3$ to both the test tubes.
- Appearance of white turbidity will indicate the presence of common salt.

Urea

Protein supplements e.g. fish meal, groundnut cake, cotton seed cake mustard cake, soybean cake, etc. are adulterated with urea to enhance their crude protein value. Since protein is a costly ingredient thus, adulteration with urea significantly enhances the profit margin. In proximate system of analysis to arrive at the crude protein value, total nitrogen in the feed sample is estimate, which can be from protein or non-protein sources. Therefore, checking the protein ingredients for urea is important.

Principle

Urea in feed and feed ingredients is extracted by treating with zinc acetate in glacial acetic acid. (Carrez solution 1) and potassium ferrocyanide (Carrez solution 2). When the filtrate is treated with dimethylaminobenzaldehyde reagents, the intensity of yellow colour which develops is directly proportional to the concentration of urea.

Materials required

Dimethylaminobenzaldehyde, Carrez solution 1, Carrez solution 2, activated charcoal, urea as standard, beaker, volumetric flask, pipette and test tubes.

Procedure

Take 1 g finely ground sample into a 500 ml volumetric flask. Add 1 g activated charcoal, 250 ml Carrez solution 1 and 5 ml Carrez solution 2. Shake and dilute it to 500 ml water and keep it for half an hour. Take 5 ml filtrate in a test tube and add 5 ml DMAB solution into it and shake. The yellow colour intensity will directly reflect the amount of urea present in the test tube.

Alternate test

Apparatus: Beaker, test tube, stirring rod, filter paper, funnel

Reagents

a. 1.6% 2,4 Dimethyl amino benzaldehyde (DMAB): Take 1.6 g of 2,4 DMAB and add 90 ml methanol+ 10 ml conc. HCl

b. Methanol

c. Conc. HCl

Procedure

1. Take 2-3 g of feed sample in a beaker.
2. Add about 40 ml water and stir the contents to dissolve urea, if present.
3. Filter the contents through Whatman filter paper no. 1.
4. Take 1 ml of filtrate in a test tube.
5. Take 1 ml distilled water in another test tube.
6. Add about 2 ml of 2,4 DMAB solution to both the test tubes.
7. Appearance of deep yellow colour indicates the presence of urea in the sample.

Heavy metals

Sr. No.	Metal	Test solutions		
		Ammonium carbonate (10%)	Ammonium sulphide (20%)	Potassium iodide (1%)
1	Antimony	White	Orange	No reaction
2	Arsenic	No reaction	Yellow	No reaction
3	Barium	White	No reaction	No reaction
4	Bismuth	White	Brown	Brown
5	Copper	Light blue	Black	Brown
6	Lead	White	Black	Golden yellow
7	Mercury	White	Black	Green terns to red
8	Zinc	White	White	No reaction

Procedure

Take a small amount of metal salt or the test material in a test tube and add about 10 ml distilled water. Then take 3 drops each of above reagents at different places on a marble slab. Put two drops of metal solution over these spots. Mix them properly and observe the colour formed. Match it with the chart and detect the metal.

Castor husk

Principle

The method is based on the fact that castor husk is not bleached under the conditions which cause the bleaching of almost all other materials of vegetable origin likely to be present in an oil cake. The method consists of the treatment of the material with dilute alkali and acid solutions followed by treatment with bleaching powder solution and isolation of the unbleached castor husk.

Materials required

White photographic dish/enameled tray, sodium hydroxide (5% w/v), hydrochloric acid solution (5% w/v), bleaching powder (5% w/v freshly prepared).

Procedure

Take three separate 100 g portion of the material and boil for 30 minutes with one litre of sodium hydroxide solution. Filter through muslin cloth, boil again for 30 minutes with one litre of hydrochloric acid solution and filter. Digest the residue for a period depending upon the type of the oil cake, with the solution of bleaching powder. When bleaching is complete, filter off the solution, spread the bleached residue in a thin layer, under water in a white photographic tray. Remove any black pieces and examine microscopically. After identification, compare the pieces with portion of castor husk, which have also undergone the above treatment (control). Castor husk has a distinctive pitted surface, when examined in reflected light under a microscope.

Mahua cake

Principle

The method is based on the principle that the saponin gives a typical colour test when extracted and reacted with antimony trichloride.

Materials required

Extraction tube (150 mm × 13 mm with a tapered tip having internal diameter of 1.5 mm), antimony trichloride solution (dissolve 125 g antimony trichloride in 300 to 400 ml of chloroform). Add 5 g calcium chloride and filter while hot. Dilute the filtrate to 500 ml with chloroform) and rectified spirit (95% v/v).

Procedure

Take 10 g of the finely ground material in the extraction thimble. Tap it to pack it well and pour rectified spirit into the tube so as to soak the sample. Collect the first drop of extract on a Whatman filter paper 1 of about 10 cm diameters. Dry and then wash it with 2 to 3 drops of distilled water in the centre of the dried spot. Dry the filter paper completely. Dip the paper in a beaker containing antimony trichloride solution and dry it on a spirit lamp or a burner. Care should be taken not to overheat the paper which will be evident by its charring. Appearance of a pink colour after heating for five minutes indicates the presence of mahua cake.

Detection undesirable constituents in feed

Hydrocyanic acid

The method is based on the release of HCN from cyanogenic-glucosides through the action of an endogenous enzyme in the fresh green samples which reduces sodium picrate compound to reddish compound. The amount of this compound is proportionate to the amount of HCN released.

Materials required

Picric acid, sodium carbonate and Whatman filter paper no. 1.

Procedure

Prepare a 5 cm × 2 cm size Whatman filter paper No 1 and dip it in 2% solution of picric acid. Dry it in room temperature and then dip it in 0.5% sodium carbonate solution and dry it again. Place about a gram of crushed test material in a test tube and add 5 to 8 drops of chloroform or toluene. Suspend the filter paper strip into the test tube and hold it with the help of a rubber cork and paper pin. Incubate it at about 38 to 40^0C overnight. A change of colour or strip from yellow to reddish brown indicates the enzymatic release (by glucosidase of plant materials) of hydrocyanide from the cyanogens present in the leaves. The intensity of colour change depends on the amount of cyanogens present in the sample.

Nitrates/ Nitrites

Take a glass slide. Put 3 to 4 drops of 2% diphenylamine solution (prepared in conc. H_2SO_4) on one side of slide. Add 1 to 2 drops of the test materials just by the side of diphenylamine solution. When they get mixed, conversion of blue colour indicates the presence of nitrates/ nitrites.

Confirmation of nitrites

Take a small amount of test material on a slide. Place 1 to 2 drops of NEDD solution (0.02%, N-1-Napthyl ethylene diamine dihydrochloride). Formation of pink colour indicates the presence of nitrites.

Confirmation of nitrates

Take 1 ml of test materials on a slide. Place 3 to 4 ml of 5% salicylic acid (in conc. H_2SO_4) solution. Change in colour from yellow to brown indicates the presence of nitrates.

Tannins

Principle

The method is based on the principle that secondary metabolites of tannins give a typical colour test when extracted and reacted with vanillin.

Materials required

Vanillin, methanol, hydrochloric acid and filter paper.

Procedure

Crush some fresh leaves and moisten the filter paper with their extract. Dip the filter paper in a solution of vanillin prepared in methanol (4 g vanillin in 100 ml methanol) for 15 seconds and then for 5 seconds in 1 N hydrochloric acid. Appearance of pink colour in the filter paper, where the leaf extract was placed, indicates the presence of tannins. The development of pink colour may take 15 minutes to 1 hour depending upon the amount of tannin present.

Alkaloids

Principle

The method is based on the principle that alkaloid gives a typical colour when extracted and reacted with dragendorff reagent.

Materials required

Ethanol, chloroform, dragendorff reagent (0.06 g bismuth subnitrate in 2 ml conc. HCl, potassium iodide, ammonium hydroxide, acetic acid, test tube and warring blender or pestle and mortar.

Procedure

Prepare an extract of about 2 g fresh leaves or feed in about 10 ml methanol or ethanol (containing 10% acetic acid). To the sample extract add dilute (50:50) ammonium hydroxide in small amounts. Formation of precipitate is an indication of the presence of alkaloids. Now dissolve the precipitate obtained in ethanol or methanol. To it add a) drop of dragendorff reagent and 10 ml water and b) 6 g potassium iodide in 10 ml water. Mix a and b and then add 7 ml conc. hydrochloric acid and add 15 ml water. Alkaloid will produce a typical yellowish brown colour.

Saponins

Saponins have a significant characteristic property of foaming in aquous solution.

Materials required

Test tube

Procedure

1. Prepare an extract of the leaves or feed in 50% aquous methanol or ethanol (or in water if methanol or ethanol is not valid). Shake the extract in a test tube. Formation of persistent foam on the liquid surface is indicative of the presence of saponins.
2. Suspend red blood cells (RBC prepared from blood of any animal available) in normal saline (0.9% NaCl solution). Add to it a small amount of aqueous extract of the feed or leaves. Haemolysis of RBC shows the presence of saponins.

CHAPTER 17

DEMONSTRATION OF CONDUCTING DIGESTION AND METABOLISM TRIAL IN SMALL RUMINANTS

Laboratory based (proximate) analysis provide quantitative information on the nutrient composition of a feedstuff; however, do not provide consistent information on the efficiency of nutrient utilization by the animal. To provide more information on the nutritional value of a feedstuff, laboratory-based analyses are used in conjunction with animal based analyses such as feeding, digestion, and balance trials. Feeding trials are preliminary trials and will be followed with additional trials, such as digestion and balance trials (5-7 days), to more specifically estimate nutrient utilization.

Digestion trial

Digestion trials provide an estimate of the total amount of a specific nutrient absorbed or utilized from the gastrointestinal tract. Digestion trials can be performed to measure the digestibility of a range of feed components including the general nutrient class or a specific component within the class. The two measures of digestibility are apparent and true digestibility. Apparent digestibility takes into account for endogenous sources (intestinal cells, enzymes, and microbes) and true digestibility does account for endogenous sources. Generally, digestion trials measure apparent digestibility. The initial step is to determine the total amount of the feedstuff consumed and the concentration of the specific nutrient in the feedstuff. The next step is to determine the total amount of feces excreted and the concentration of the specific nutrient in the feces. In digestion trial, recording of feed intake and feces voided is required. Measurements are expressed on a dry matter basis (DM basis) to correct for differences in moisture content.

Metabolism or balance trial

Balance trials are another method to determine nutrient utilization of a feedstuff. Compared to digestion trials, balance trials provide a more complete and accurate picture of nutrient utilization of a feedstuff in the animal. Balance trials attempt to estimate the net retention of a specific nutrient in the animal. Balance trials may be performed on a range of feed components. Similar to digestion trials, balance trials measure total amount of a nutrient consumed and total amount of a nutrient excreted and use an equation to measure the resultant value. In a balance trial, individual intake is measured similar to the first method in a digestion trial. However, in a balance trial the measured sources of excretion are feces, urine, respired air, and sloughed materials such as skin, hair, and feathers. For lactating animals, milk yield and components are also included as a source of excretion. To collect the required measurements, balance trials require the use of a metabolism crate. In metabolism trial, along with recording of feed intake and feces voided, recording of milk yield, wool production, etc. also required.

Digestion and metabolism trial for sheep and goat

Principle

Weighed amount of experimental feed is fed to sheep and goat for a definite period of time and the excreta resulting there from are collected quantitatively for chemical analysis. By knowing the proximate principles in feed and faeces, digestion coefficients and digestible nutrients can be calculated. If we analyze the urine and products such as mild or wool the amount of source of nutrients absorbed and retained in the body or stored as a product may be calculated by difference.

Apparatus

1. Metabolism cages with an arrangement at the floor for separation of urine and faeces
2. Urine collecting containers (simple glass bottles or plastic containers).
3. Faeces containers, plastic bags or plastic buckets.
4. Meat grinder.
5. Weighing balance (to weigh animals, feed, faeces and urine).
6. Wide mouth glass bottles.
7. Plastic bottles.

Selection and grouping of animals

The animals should be of the same breed, sex, age and body weight. Generally a minimum of four animals in each group is required. As number of animals increases error will be decreases. Animals should be healthy and free from parasites. Male animals are preferred to females because it is easier to collect faeces and urine separately.

Feeding and watering

Experimental sheep and goat should be fed at the same time each day early in the morning at about 9 A.M. The amount of experimental ration should be exactly weighed. Experimental sheep and goat should be allowed to drink clean and wholesome water at least three times a day in winter season and six times in summer season. If mineral balance studies are to be conducted then their contents should also be analyzed in water. About one litre of water (representative sample) should be evaporated in a clean double distilled water washed glass beaker. The volume of water drunk by sheep should be recorded during the period of metabolism trial. Always de-worm the experimental sheep before starting preliminary period. Experimental sheep should be healthy in all respect so as to study the normal nutrient metabolism.

Preliminary Period

The purpose is to make the gastrointestinal tract free of any indigestible material from the feed consumed prior to the start of the digestion trial. The main purpose of the preliminary period is to acquaint the animal with the metabolism cages or stall, to make the necessary adjustments in order that the faeces and urine are collected properly and to adjust the animal to its intake of feed in relation to the excretion of faeces and urine. In monogastric animals such as pigs, digestion and evacuation are usually complete in 24 hours after the test feed is ingested. So, preliminary period of 2-3 days is followed. In ruminants eating a roughage ration, the last residues may not be voided until 150 to 200 hours (6 to 8 days). So, preliminary period of 7-14 days (8 to 10 days) is followed to eliminate the feed residues of previous rations from the digestive tract and to stabilize the daily feed intake and faecal output to a constant level.

Collection Period

1. Weigh the experimental sheep and goat at 7 A.M. before offering water and experimental ration, on three consecutive days. Take average weight of three day's recording for consideration in the calculation of digestion trial data. Similarly, take the weight of experimental sheep after the completion of collection period.
2. Animals are transferred to cages or stalls and a 2-3 day adaptation period is allowed for their acclimatisation.
3. Usually collection period of 7-10 days duration was followed; a 5-7 day collection period is mostly followed. If the animal is on *ad libitum* feeding, the collection period should be at least 14 days long. The collection period should begin on the morning after the animal has been eating a constant amount of feed.
4. In general, the longer the period of collection the more accurate the results are, provided the stated amount of feed continues to be consumed regularly and completely, since the effect of periodic fluctuations is minimised.

5. Quantitative collection of faeces and urine are done.
6. Representative samples of feed offered, feed leftovers and faeces and urine voided are preserved and analysed for nutrient composition.

Sampling and processing

Samples of feeds and fodders offered and residue left, faeces voided, urine excreted etc should be processed. Faeces and urine voided during 24 h was recorded for seven days and daily representative sample was brought to laboratory for aliquoting. A known amount (1/100th part) of representative faecal sample was taken in pre-weighed dried aluminium tray for determination of dry matter in faeces. After determination of dry matter, dried faecal sample for each animal was pooled for 7 days. For N determination, 1/100th of the total faeces voided were kept in a pre weighed plastic container and known quantity of 25% H_2SO_4 was added. Similarly, 1/100 parts of the total urine voided was kept in separate bottle and known quantity of 25% H_2SO_4 was added. For estimation of minerals, a known amount (1/100th parts) was taken in clean glass bottle and toluene was added as preservative. If gross energy of urine sample is to be determined then take aliquot 1/100th in a screw capped polyethylene bottle and keep sample in refrigerator at 1 °C. At the end of collection period mix thoroughly the sample of urine and transfer 20 ml measured volume in the Kjeldahl flash so as to estimate total nitrogen.

Calculation

$$\text{Digestibility coefficient (\%)} = \frac{\text{Nutrient intake} - \text{Nutrient outgo}}{\text{Nutrient intake}} \times 100$$

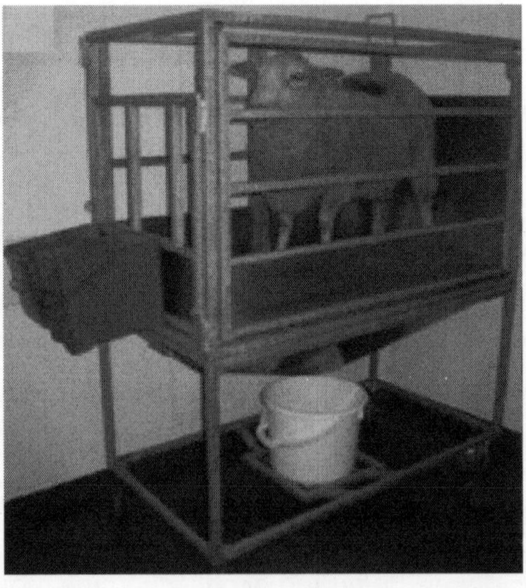

Figure 1. Conducting metabolism trial in sheep

CHAPTER 18

DEMONSTRATION OF CONDUCTING DIGESTION AND METABOLISM TRIAL IN LARGE RUMINANTS

Procedure for conducting digestion and metabolism trial is same as mentioned in the case of small ruminants except the following mentioned points:

1. In the case of male animals (bullocks), use of urine bags are advised, while faeces is collected manually and kept in the plastic buckets.
2. In the case of female animals (cows and buffalos), polyethylene tube (catheter) fitted in the urethral orifice and supported by adhesive tap fixed on the labia major, is used for collection of urine.
3. Aliquot of urine (1/1000) is taken directly in to the Kjeldahl flask for nitrogen estimation: another aliquot of urine (1/1000) is taken clean polyethylene bottle and kept in the refrigerator at 1°C for the estimation of calorific value. Aliquot of faces (1/400) is taken and preserved with dilute sulphuric acid (30%) for nitrogen estimation. Another aliquot of faeces (1/200) is taken for dry matter estimation. Aliquot of milk (100 ml per litre of milk produced) is taken at each milking; samples of three times milking of every day are pooled up and kept in polyethylene bottles at 4°C in a refrigerator. Another aliquot (1/1000) out of this times pooled samples is taken in duplicate, one is used for nitrogen estimation and another is kept at 1°C in refrigerator for the estimation of gross energy. Seven days samples of urine specified for gross energy estimation is also kept in refrigerator at 4°C.

Aliquoting of faeces and urine samples

Daily amount of faeces voided is recorded and 1/50th total weight is taken for dry matter estimation and preserved daily in paper bags. Another aliquot of faeces 1/100 is carefully weighed in a water glass and transferred to a glass mortar and mixed with 5 ml 30% sulphuric acid for mixing the nitrogen, the mixed sample is taken in wide mouth glass bottle. Amount of faeces collected in the bottle during a period of five days is weighed and weight of sulphuric acid added is deducted from the total weight so as to know the actual weight of faeces in the bottle. Weigh 20 g faeces on the filter paper and transfer to Kjeldahl flask for digestion and estimating total nitrogen by macro-Kjeldahl method. The sample of urine is carefully aliquoted using standard graduated pipette. An aliquot 1/100 is taken in the Kjeldahl flask for nitrogen estimation and duplicate aliquot 1/100 is taken in a clean capped bottle, mixed with 1 ml sulphuric acid. The cumulative sample of five days period is collected in the Kjeldahl flash and then digested as such. If copper or other trace element in the urine is to be determined then another aliquot 1/100 is taken in the glass bottles rinsed with glass distilled water and 1 ml of distilled sulphuric acid is added to each bottle so as to prevent fermentation.

CHAPTER 19

CALCULATION OF NUTRITIVE VALUE/DIGESTIBILITY OF DIFFERENT FEEDSTUFFS BY DIRECT METHOD

Objective

To determine dry matter digestibility, crude fibre digestibility, ether extract digestibility, crude protein digestibility, nitrogen free extract digestibility, total digestible nutrients (TDN) and nutritive ratio (NR).

Estimation of digestibility coefficient in ruminant by direct method

Digestibility determination by direct method is done by following digestion or metabolism trials.

Materials required

- Metabolic stall (or metabolic cages)
- Experimental animals
- Faeces and urine collection bags
- Feed sample etc.

Procedure

As mentioned in practical 1 and 2.

Observations

1. Feed intake (feed offered - feed left over)
2. Dry matter intake

3. Dry matter voided in faeces
4. Weight of empty faecal bag
5. Weight of faecal bag + faecal sample
6. Weight of faecal sample

Calculation

$$\text{Digestibility coefficient (\%)} = \frac{\text{Nutrient intake} - \text{Nutrient outgo}}{\text{Nutrient intake}} \times 100$$

Factors influencing the digestibility coefficient

The following factors influence the digestibility co-efficient of various nutrients from the same feed stuffs:

1. Species of the animal
2. Age of the animal
3. Level of feeding
4. Processing of feeds and fodders
5. Feed composition
6. Ration composition

Limitations

There are two main limitations in interpretation of digestibility co-efficients. The first is that the faeces should represent the undigested fraction of food only for which the digestibility co-efficient is being determined. But part of the nutrients excreted in the faeces come from the unspent enzymes, cellular material abraded from the gut and other secretions of the gut which actually represent the utilized nutrients of previous diets. The second limitation of digestibility trial is that the carbohydrates are broken down to volatile fatty acids, CO_2 and CH_4. The latter two do not give the energy to the animals but are computed as digestible carbohydrate since they are not recovered in faeces. This loss leads to over estimation of digestible carbohydrates.

Calculation of digestible nutrients

$$\text{Percent digestibility nutrient} = \frac{\text{\% nutrient in feed} \times \text{Digestibility \% of nutrient}}{100}$$

Therefore, for calculation of DCP %

$$\text{DCP \%} = \frac{\text{\% CP in feed} \times \text{Digestibility \% of CP}}{100}$$

Similarly, the DCF%, DNFE% and DEE% are calculated to derive the TDN value of feed.

Calculation of TDN% on dry matter basis

The formula for calculating TDN% is as follows:

TDN% = DCP% + DCF% + DNFE% + (DEE% × 2.25)

The digestible fat is multiplied by the factor 2.25. This is the correction factor for arriving at digestible carbohydrate equivalent from digestible fat according to Atwater fuel values (carbohydrate: 4 Kcal/g; fat 9 Kcal/g) because the values have taken account of only digestible losses in case of these nutrients. TDN value of a feedstuff can be used to derive other form of energy by using the following relationship.

1 Kg TDN= 4409 Kcal DE or 3616 Kcal ME

Calculation of Nutritive ratio (NR)

NR of a particular feed may be defined as the ratio between the digestible protein and digestible non-nitrogenous nutrients (%digestible carbohydrate + %digestible ether extract × 2.25).

$$NR = \frac{\%\text{ digestible carbohydrate} + (\%\text{ digestible EE} \times 2.25)}{\%\text{ digestible CP}}$$

$$NR = \frac{\text{TDN \%} - \text{DCP \%}}{\text{DCP \%}}$$

Example 1. Calculate the DCP, TDN and NR in feed by conventional method when a group of bullocks were fed on an average 4.00 kg hay/head/day. On average 6.00 kg faeces were voided. Chemical composition on fresh basis (%) is given below:

Nutrients	Hay	Silage	Faeces
Moisture	16.00	65.00	75.00
Ash	6.00	3.88	2.00
Crude protein	12.57	2.80	3.22
Crude fibre	27.78	12.57	9.50
Nitrogen free extract (NFE)	35.15	14.56	9.74
Ether extract	2.50	1.19	0.54

Solution:

A. Calculation of TDN% in a feed

a) Convert the chemical composition (%) on DM basis

Nutrients	Hay	Silage	Faeces
Crude protein	14.96	8.00	12.88
Crude fibre	33.07	35.9	38.00
Nitrogen free extract	41.85	41.6	38.96
Ether extract	2.98	3.40	2.16
Ash	7.14	11.10	8.00

b) DM of hay intake and faeces outgo/day

DM of hay = 3360.00 g, DM of faeces = 1500.00 g

c) Calculation of nutrient intake and nutrient outgo except ash

Nutrients	Intake from hay (g)	Outgo from faeces (g)
Crude protein	502.66	193.20
Crude fibre	1111.15	570.00
Nitrogen free extract	1406.16	584.40
Ether extract	100.13	32.40

d) Calculation of digestibility coefficients of dry matter and other nutrients by applying the formula

$$\text{Digestibility coefficient} = \frac{\text{Nutrient intake} - \text{Nutrient outgo}}{\text{Nutrient intake}} \times 100$$

$$\text{DM digestibility (\%)} = \frac{3360 - 1500}{3360} \times 100 = 55.36\ \%$$

$$\text{CP digestibility (\%)} = \frac{502.66 - 193.20}{502.66} \times 100 = 61.56\ \%$$

$$\text{CF digestibility (\%)} = \frac{1111.15 - 570.00}{1111.15} \times 100 = 48.70\ \%$$

$$\text{EE digestibility (\%)} = \frac{100.13 - 32.40}{100.13} \times 100 = 67.64\ \%$$

$$\text{NFE digestibility (\%)} = \frac{1406.16 - 584.40}{1406.16} \times 100 = 58.44\ \%$$

e) Calculation of % digestible nutrients

Percent digestible nutrient

$$= \frac{\text{\% nutrient in feed} \times \text{Digestibility \% of nutrient}}{100}$$

$$\text{DCP\%} = \frac{14.97 \times 61.56}{100} = 9.21\ \%$$

$$\text{DCF\%} = \frac{33.07 \times 48.70}{100} = 16.11\ \%$$

$$\text{DEE\%} = \frac{2.98 \times 67.64}{100} = 2.02\ \%$$

$$\text{DNFE\%} = \frac{41.85 \times 58.44}{100} = 24.46\ \%$$

f) Calculation of TDN% on dry matter basis (DMB)

TDN % = % DCP + % DCF + % DNFE + (2.25 × % DEE)

= 9.21 + 16.11 + 24.46 + (2.25 × 2.02)

= 9.21 + 16.11 + 24.46 + 4.55

= 54.33 %

B. Calculation of nutritive ratio (NR)

$$\text{NR} = \frac{\text{TDN \%} - \text{DCP \%}}{\text{DCP \%}}$$

$$= \frac{54.33 - 9.21}{9.21}$$

= 4.90

= 4.90:1

Answer: DCP = 9.21%; TDN = 54.33%; NR = 4.90:1

CHAPTER 20

CALCULATION OF NUTRITIVE VALUE/DIGESTIBILITY OF DIFFERENT FEEDSTUFFS BY INDIRECT METHOD

There are two methods of determination of digestibility by indirect method:

1. Indicator method
2. Difference method

Indicator method

Determination of digestibility by indicator method.

Characteristics of an indicator

Inert reference substance is generally used as indicator.

1. It should be totally indigestible.
2. Un-absorbable in the gut.
3. Have no pharmacological action on digestive tract.
4. Pass through digestive tract at uniform rate.
5. Be readily determined chemically.
6. Preferably be a natural constituent of feed under list.

Types of indicators

Indicators are of two types

1. Internal/natural e.g. lignin, chromogens, acid insoluble ash.
2. External indicator e.g. ferric oxide, chromic oxide, etc.

Procedure

In the conventional method of determining the digestibility coefficients we have to do a quantitative collection of daily feed intake and faeces voided. This can be avoided in some cases by the use of index substances or reference substance. By determining the ratio of the concentration of the reference substances to that of the given nutrient in the feed and faeces, indigestibility of the nutrient is obtained. The calculations for the digestibility are made from the following formula:

$$100 - \text{Indigestibility} = \text{digestibility}$$

$$\text{Digestibility} = 100 - \left(100 \times \frac{\%\text{ of indicator in feed}}{\%\text{ of indicator in faeces}} \times \frac{\%\text{ of nutrient in faeces}}{\%\text{ of nutrient in feed}}\right)$$

Example 1. A pig was given 4.50 kg feed (10% moisture) per day and 1.0 g of indicator was mixed in feed before feeding. The faeces of pig were collected upto 24 h and the estimated indicator in faeces was 0.5 mg/ g of DM of faeces. Calculate the DM of faeces and digestibility of DM and other nutrients. The chemical composition of feed and feces were as under (on %DM basis):

Nutrient	Feed (%)	Faeces (%)
Crude protein	23.00	20.00
Crude fibre	10.00	12.00
Nitrogen free extract	52.00	55.00
Ash	10.00	8.00
Ether extract	5.00	5.00

Calculation of fecal DM output

$$\text{Faecal DM (g)} = \frac{\text{Indicator intake (mg/day on DM basis)}}{\text{Indicator concentration in feces (mg/g on DM basis)}}$$

Therefore, Faecal DM output = 1000 mg 0.5 mg/g = 2000 g

Calculation of indigestibility of dry matter

Indigestibility of DM (%) = 100 - % digestibility of DM

$$\text{Indicator (\%) in feed on DM basis} = \frac{\text{Indicator (g) in feed used}}{\text{Feed intake (g) on DM basis}}$$

Digestibility of dry matter (%) = 100 – Indigestibility (%) =100 – 49.30 = 50.70%

Calculation of the digestibility of nutrients

$$\text{Digestibility} = 100 - \left(100 \times \frac{\%\text{ of indicator in feed}}{\%\text{ of indicator in faeces}} \times \frac{\%\text{ of nutrient in faeces}}{\%\text{ of nutrient in feed}}\right)$$

Digestibility of CP (%)

$$= 100 - \left(100 \times \frac{0.0246}{0.05} \times \frac{20}{23}\right)$$

= 100- 42.7

= 57.25 %

Digestibility of CF (%)

$$= 100 - \left(100 \times \frac{0.0246}{0.05} \times \frac{12}{10}\right)$$

= 40.96 %

Digestibility of NFE (%)

$$= 100 - \left(100 \times \frac{0.0246}{0.05} \times \frac{55}{52}\right)$$

= 48.34%

Digestibility of EE (%)

$$= 100 - \left(100 \times \frac{0.0246}{0.05} \times \frac{5}{5}\right)$$

= 50.80%

CHAPTER 21

CALCULATION OF NUTRITIVE VALUE/DIGESTIBILITY OF DIFFERENT FEEDSTUFFS BY DIFFERENCE METHOD

Digestibility of individual feed may be determined if it provides a satisfactory ratio for the test period when fed alone like green maize, oat, grasses, legumes etc. Some feeds however cannot be fed alone and therefore their digestibility co-efficient cannot be determined directly like concentrates such as maize grain, cakes and bran as they do not supply the bulk. In case of non maintenance type of roughages like straws and stovers, also, direct digestibility determination cannot be made. Since during test-period, the feed in question does not supply the required nutrients.

In this indirect system of digestibility determination, two or more digestibility trials are conducted. In the first trial, nutrients are determined by feeding a basal maintenance type of fodder where the feed in question is not fed. For instance, if the digestibility of nutrients of groundnut cake is to be estimated, then first trial may be given with oat hay. In this trial, digestibility of nutrient present in the oat hay will be determined. In the 2nd trial, oat hay + groundnut cake will be fed. In this case it will assume that the digestibility coefficients of the nutrients in the oat hay will remain the same in the 2nd test as was determined in the 1st trial. Therefore from the total nutrients voided in the faeces the portion belonging to oat hay is subtracted. The remaining faecal nutrients are considered to have come from the ground nut cake or the food under test.

For determination of crude protein digestibility, of the cake, similar calculations are made. These calculations may not be very correct since the addition of cake to the basal diet may influence the digestibility of the basal diet, but the credit is given to the supplemented diet assuming that the digestibility would remain the same as if basal diet tested alone. This effect is known as associative effect of feeds.

In case of determining the digestibility coefficients of wheat straw, three sets of trials are conducted. First trial is conducted with maintenance quality fodder, then fodder + cake and third trial with cake + straw. Since the straw is deficient in protein, it cannot be fed alone to the animals. For determining the digestibility coefficient of the straws, cakes of known digestibility coefficients are fed. In all these trials the same basic assumption is made that digestibility of the nutrients in the basal diet will remain the same when fed alone or with supplements.

Example 1. Find out the DCP and TDN of maize grain when fed to the bullock along with hay. The bullock received 4.0 kg of hay/ day along with 4.20 kg of ground maize. The average daily excretion of faeces on hay alone and mixed ration was 5.0 kg and 8.50 kg, respectively. Composition of hay, maize and faeces (fresh basis) is as follows:

Nutrient	Hay	Maize	Faeces when hay was fed	Faeces when hay + maize were fed
Moisture	16.00	12.57	75.00	80.95
Crude protein	12.57	9.65	3.22	3.55
Crude fibre	27.78	1.90	9.50	6.50
Nitrogen free extract	35.15	70.85	9.74	6.75
Ether extract	2.50	3.65	0.54	0.50
ash	6.00	1.20	2.00	1.75

Solution:

a) Convert the chemical composition on dry matter basis

Nutrient	Hay	Maize	Faeces when hay was fed	Faeces when hay + maize were fed
Crude protein	14.96	11.06	12.88	18.63
Crude fibre	33.07	2.18	38.00	34.12
Nitrogen free extract	41.85	81.20	38.96	35.43
Ether extract	2.98	4.18	2.16	2.62
ash	7.14	1.37	8.00	9.18

b) DMI/ day through hay, maize, hay + maize and faecal DM outgo when only hay was fed and mixed ration (hay + maize) were fed:

DM intake through hay = 3360.00 g

DM intake through maize = 3664.50 g

Total DM intake = 3360 + 3664.50 = 7024.50 g

Faecal DM when hay was fed alone = 1450.00 g

Faecal DM when hay + maize were fed = 619.25 g

c) Calculation of total nutrient intake, total nutrients outgo and total nutrients digested (g/day) except ash

Nutrients	Nutrients intake from hay	Nutrients intake from maize	Total nutrients intake from mixed ration	Nutrient outgo when hay was fed	Nutrient outgo when hay mixed ration was fed	Nutrient digested from hay	Total nutrient digested when mixed ration was fed
	a	b	a +b	c	d	a-c	(a+b)-d
CP	502.66	405.30	907.96	186.76	301.67	315.90	606.29
CF	1111.15	79.89	1191.04	551.00	552.49	560.15	638.55
NFE	1406.16	2975.57	4381.73	564.92	573.70	841.24	3808.03
EE	100.13	153.18	253.31	31.32	42.42	68.81	210.89
DM	3360.0	3664.50	7024.50	1450.00	1619.25	1910.00	5405.25

d) Calculation of the nutrients and the dry matter digested from maize by subtracting estimated nutrients and dry matter digested from hay from total digested nutrient and dry matter (g).

Nutrient	Total digested	Estimated digested from the hay	Calculated digested from maize
CP	606.29	315.90	290.39
CF	638.55	560.15	78.40
NFE	3808.03	841.24	2966.79
EE	210.89	68.81	142.08
DM	5405.25	1910.00	3495.25

e) Calculation of digestibility coefficient of DM and other nutrients of maize by applying the formulae

$$\text{Digestibility coefficient} = \frac{\text{Intake} - \text{outgo}}{\text{Intake}} \times 100$$

$$= \frac{\text{Digested}}{\text{Intake}} \times 100$$

$$\text{Digestibility coefficient of CP} = \frac{290.39}{405.30} \times 100 = 71.65\%$$

$$\text{Digestibility coefficient of CF} = \frac{78.40}{79.89} \times 100 = 98.13\%$$

$$\text{Digestibility coefficient of NFE} = \frac{2966.79}{2975.57} \times 100 = 99.70\%$$

$$\text{Digestibility coefficient of EE} = \frac{142.08}{153.18} \times 100 = 92.75\%$$

$$\text{DM digestibility} = \frac{3495.25}{3664.50} \times 100 = 95.38\%$$

f) Calculation of digestible nutrients of maize

$$\%\ \text{Digestibility nutrients} = \frac{\%\ \text{of nutrient in feed} \times \text{digestibility}\ \%\ \text{of nutrient}}{100}$$

So, DCP % = 11.06 x 71.65/ 100 = 7.92%

DCF % = 2.18 x 98.13/ 100 = 2.14%

DNFE % = 81.20 x 99.70/ 100 = 80.95%

DEE % = 4.18 x 92.75/ 100 = 2.25%

g) Calculation of TDN % of maize

TDN % = DCP% + DCF% + DNFE% + (%DEE × 2.25)

= 7.92 + 2.14 + 80.95 + (3.88 × 2.25)

= 7.92 + 2.14 + 80.95 + 8.73

= 99.74%

$$\text{TDN \% (fresh basis)} = \frac{99.74 \times 87.25}{100} = 87.02\%$$

DCP % (DMB) = 7.92%

$$\text{DCP \% (fresh basis)} = \frac{7.92 \times 87.25}{100} = 6.91\%$$

CHAPTER 22

GENERAL OUTLINES FOR RATION FORMULATION OF RUMINANTS

The formulation of ration for various categories of livestock depends upon the nutrient requirements for a particular animal or group of animals and the feeds and fodders available for the purpose. The nutrient requirements are largely based on the size of the animal and its physiological state i.e. growth, pregnancy, milk production, etc. The livestock feeds can be grouped into two major categories i.e. roughages and concentrates on the basic of bulkiness and chemical composition.

Roughages

These are bulky feeds containing more than 18% CF and less than 60% TDN. Roughages are sub-divided into two major groups-dry (>80% DM) and succulent (80% DM). The dry roughages are hays (leguminous and non-leguminous), straw (wheat straw, paddy straw, barley, straw, etc.) and stovers (bajra, jowar, maize, etc.). Except good quality leguminous and non-leguminous hays, remaining are non-maintenance type roughages. Succulent roughages include mainly cultivated leguminous and non-leguminous fodders. The most commonly used kharif leguminous fodders are cowpea and guar and non-leguminous ones are jowar, maize and sudan grass. The leguminous and non-leguminous fodders of rabi season are Lucerne, berseem, pea and oat and barley respectively. In addition, pasture, tree leaves silages and root crops are also grouped under this head.

Concentrates

Feed or feed mixtures containing less than 18% CF and over 60% TDN are classed as concentrates. They are rich in either energy or protein and accordingly are called as energy concentrates (CP < 20%) or protein concentrates (CP > 20%). Various energy concentrates are grain and seeds (maize, barley, sorghum, etc.), milling byproducts (brans and chunis), roots (tapioca tubers turnip, potatoes) and molasses. Protein supplements, on the other hand may be further categorized according to the source of origin (a) animal protein supplements (meat meal, blood meal, fish meal, feather meal, etc.) and (b) plant protein supplements (oilseed cakes and meals i.e. groundnut cake, sesame cake, mustard cake, linseed oil meal, etc.). Animal origin protein supplements except dried whole and skimmed milk contain more than 47% CP, whereas those of plant origin, contain less than 47% CP except soybean oil meal (CP 49%). In addition, a third group of feedstuffs include mineral and vitamin supplements and additives (antibiotics, hormones, etc.) which when added to the total feed mixture improve the performance of the total feed.

Formulation of rations

The ration of animal may be divided into two parts viz. for maintenance and for production purpose.

1. Maintenance ration

The ration which just enables the animal at rest to carry on the essential processes of life such as breathing, circulation of blood etc without either gain or loss of body weight is known as maintenance ration. It is dependent upon live weight of animal and also on the level of production. At higher level of production even at the same body weight, the maintenance requirement is increased.

2. Production ration

It is the additional allowance of ration for milk product over and above the maintenance requirement. An additional amount of 1 kg concentrate mixture is required for each 2.5-3 kg milk produced. Production also includes growth, fattening work etc. The additional allowance will depend upon rate of growth, fattening and type of work being done.

3. Gestation ration

It is the feed allowance which should be given to pregnant animal from fifth-sixth month of pregnancy onwards for proper growth of the fetus and to keep the mother fit for optimum milk production upon calving. An additional amount of 1.25 and 1.75 kg of concentrate mixture has been recommended for cows/ buffaloes, respectively.

4. Balanced ration

A balanced ration is one that furnishes nutrients in such proportions and amounts that will properly nourish a given animal for 24 h. In addition, the required nutrients must be contained in the amount of DM the animal is able to consume in the 24 h period.

Steps of ration formulation

1. To calculate the probable DMI of the animal.
2. To calculate the nutrient requirement of the animal
3. To arrive at the amount of available feeds that must be fed to fulfill the animal's nutrient requirement within its expected DMI.

General guidelines for calculation of DMI

Dry matter intake (DMI) plays an important role in nutrition as it determines the quantity of nutrients accessible to an animal for its well-being and productivity.

Physical fill of the reticulorumen (Allen, 1996; Mertens, 1994)

Equations for predicting DMI

1. Lactating Cows

The 1971 edition of the National Research Council (NRC) suggested unrestricted feeding during the initial 6 to 8 weeks of lactation, followed by feeding according to energy needs for nursing dairy cows. In 1978 (NRC, 1978) body weight and 4 percent fat-corrected milk were factors used to estimate DMI, which ranged from 2 to 4 percent of body weight.

The 1989 edition of the NRC (National Research Council) made predictions about DMI based on the concept of energy requirements.

$$\text{DMI (kg)} = \frac{\text{NEL required (Mcal)}}{\text{NEL concentration of diet (Mcal/kg)}}$$

The NEL incorporated criteria pertaining to upkeep, milk production, and restoration of lost body mass. Proposed an 18% decrease in dry matter intake (DMI) within the initial three weeks of lactation. The Dry Matter Intake (DMI) decreases by 0.02 kg for every 1 percent rise in moisture content of the diet above 50%, when fermented feeds are being consumed, per 100 kg of body weight. The 2001 edition of the NRC released an equation based on data from the Journal of Dairy Science (1988-1998), as well as data from Ohio State University and the University of Minnesota (May, 1994). The equation for DMI (kg/d) is calculated as follows:

DMI = (0.372 X FCM + 0.0968 X BW^(0.75)) X (1-e^(-0.192 X (WOL+3.67)))

For temperatures over 20° C, DMI is multiplied by (1 - ((°C - 20) X 0.005922))

For temperatures below 5° C, DMI is divided by (1 - ((5 - °C) X 0.004644)).

(WOL = week of lactation).

2. Growing Heifers

DMI (kg/d) = ([X (0.2435 X NEM - 0.0466 X 0.1128)/NEM])

NEM is net energy of diet for maintenance (Mcal/kg).

The adjustment factor for above equation based on days of gestation is = [1 + ((210 - DG) X 0.0025)] where DG = day of gestation.

DM requirement varies with live weight and status/rate of the animal production. DM consumption in various Species of livestock is as follows:

Species	DM (% of live weight)
Cattle	1.5 to 2.5
Buffalo	2.0 to 3.0
Sheep	2.5 to 3.5
Goat	3.0 to 4.0 may go upto 7%

Example-Lactating dairy cows weighing an average of 500 kg will consume approximately 2.5% of their body weight in dry matter intake daily.

Dry Matter Demand (kg) = 500 x (2.5/100) =12.5 kg

To provide the required bulk in the diet, the DM allowance for cattle and buffalo should be divided as follows.

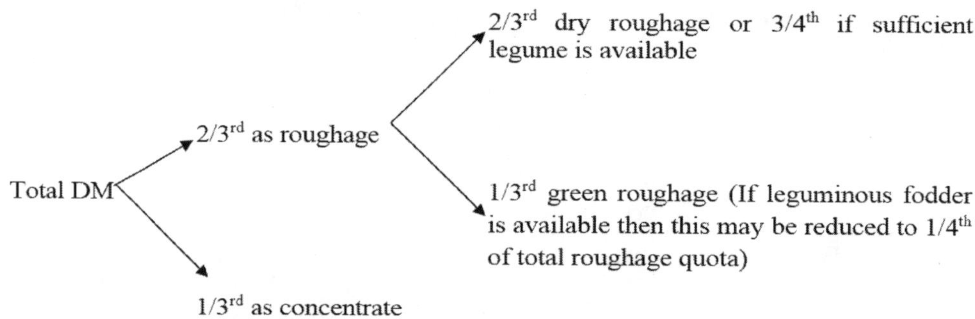

Optimal nutritional and feeding strategies for maximizing DMI in dairy Cows

Moisture:

A decrease in total DMI of 0.02 percent of body weight for each 1% increase in moisture content of the diet above 50% was shown with fermented feeds. Typically, high moisture feeds undergo fermentation, and the reduction in Dry Matter Intake (DMI) when they are consumed is believed to be primarily caused by the by-products of fermentation rather than the water content itself. Nevertheless, the soaking of grain did not have any impact on DMI. Although there is a contradiction between the dietary dry matter (DM) content and dry matter intake (DMI), there is no clear optimal DM content of the diet that maximizes DMI.

Neutral detergent Fiber (NDF):

NDF is utilized to establish the minimum and maximum limits of DMI. When the diet contains a high concentration of NDF, the amount of feed that may be consumed is limited by the amount of space in the rumen. On the other hand, when the diet contains a low concentration of NDF, the intake of energy is limited by feedback inhibitors. Research indicated that there is a consistent decrease in DMI when the content of NDF in diets increases, particularly when diets contain more than 25 percent NDF. Nevertheless, when lactating cows were provided with high energy meals containing NDF levels ranging from 25 to 42 percent of DM, less than 1 percent of the variation in DMI could be attributed to these diets.

Forage to concentrate ratio:

The DMI increased as the amount of concentrate in the diets increased, regardless of the type of forage. The digestible dry matter also showed an increase when the amount of concentrate in the meal rose. Cows that were given a diet consisting of 74 percent forage (with a ratio of 2:1 alfalfa silage to corn silage) consumed 2.7 kg less dry matter per day compared to cows that were fed a diet containing 50 percent forage. Typically, raising the amount of the concentrate in the diets up to 60 percent there is increase in the DMI.

Fat:

When fat is used as an energy source in diets, cows tend to consume less DM to meet their energy needs compared to when carbohydrates are used. Fats can also reduce ruminal fermentation and the digestion of fiber, resulting in increased rumen fill and a slower rate of passage. The inclusion of oilseeds and hydrogenated fatty acids in meals with a total fatty acid content of 5 to 6 percent had a quadratic impact on dry matter intake, with the lowest values observed at 3 and 2.3 percent additional fatty acids, respectively. As the ratio of unsaturated fatty acids in the fat source increased, the dry matter intake generally decreased. Elevating the levels of saturated fatty acids typically mitigates the adverse ruminal consequences linked to lipids. However, when fat is substituted for grain, there is an apparent increase in DMI due to a decreased heat rise during periods of heat stress and a reduction in propionate inhibition.

Feeding method—total mixed ration vs. Individual ingredient:

Nutrients can be efficiently provided by either feeding a total mixed ration (TMR) or individual ingredients. TMR enables the amalgamation of all feed components in accordance with a specified quantity of each component. Feeding animals a TMR without sorting the ingredients should result in improved rumen fermentation and more efficient utilization of nutrients compared to feeding separate ingredients.

Feeding frequency:

Studies have indicated that increasing the frequency of providing feed to cows has a positive impact on milk output and reduces the occurrence of health issues. Changing the feeding frequency from one or two feed offerings per day to four resulted in a 16 percent rise in the average daily gain of cattle, a 19 percent increase in feed utilization, a 7.3 percent increase in milk fat, and a 2.7 percent increase in milk production. The increase in productivity was most significant when cattle were provided with high-concentrate diets.

Sequence of feeding:

Feeding cows with easily fermentable carbohydrates after a period of more than 6 hours without feed might lead to acidosis conditions in the rumen, which can reduce feed intake and hinder fiber digestion. Feeding forage(s) as the first feed in the morning before other feedstuffs would allow for the formation of a fiber mat in the rumen and provide buffering capacity in the rumen from both the forage and the increased salivation associated with forage consumption.

Access to feed:

Feed access is crucial for maximizing DMI in cows. This means that cows need sufficient time for eating. There is a significant disparity in the eating habits of cows when comparing a non-competitive feeding environment to a competitive feeding environment. Restricting the cows' access to feed to only 8 hours per day resulted in a drop in milk production for cows that typically produce an average of 25 kg/day. This loss ranged from 5 to 7 percent when compared to animals that had unrestricted access to feed.

CHAPTER 23

CLASSIFICATION OF ANIMAL FEEDSTUFFS AND THEIR NUTRITIVE VALUE

Feedstuffs	Types	Example
1. Roughages >18% CF <60% TDN	**A) Dry** >80% DM	**i) Hays** (18-34% CF,40-60% TDN & Considerable carotene) Leguminous hay: (hays of Lucerne, Berseem, etc.) >10.5% CP, >0.9% Ca Non leguminous hays: (Sorghum, Doob grass, etc.) 6-10.5%, <0.9% Ca **ii) Low quality roughages** >28% CF- many >34% CF <52% TDN- several <40% TDN <6% CP, little carotene, if any (Straws like wheat, rice, oat, jowar, barley, etc.)
	B) Succulent <80% DM	**i) Green fodder** a) Leguminous (Berseem, Lucerne, Cowpea, cluster bean, green pea >10.5% CP) b) No-leguminous (Fodders of jowar, maize, bajra, oat etc. and grasses of sudan, napier, guinea, etc.) c) Pasture d) Tree leaves e) Silage (25-45% DM, medium in carotene) f) Root crops (9-30% DM, <3% CF)

Feedstuffs	Types	Example
2. Concentrates	**A) Energy rich**	**a) Grain and seeds** (Maize, barley, sorghum, etc.)
<18% CF	(<18% CP)	**b) Mill byproducts** (Wheat bran, rice bran, chunis of dals)
>60% TDN		**c) Roots** (Turnip, potatoes, tapioca, etc.)
		d) Molasses- sugarcane molasses, sugar beet, molasses, etc.
	B) Protein rich	**a) Animal origin** (mostly >47% CP, >1% Ca, >1.5% P & <2.5% CF)
	(>18% CP)	e.g. i) Animal by-products like blood meal, meat scraps, dried whole and skimmed milk.
		ii) Marine by-products
		iii) Avian byproducts
		b) Plant origin (mostly <47% CP, <1% Ca, <1.5% P & >2.5% CF)
		Defatted oil seeds (<17% EE)
		e.g. cakes and meals
		Brewer's grain and yeast (>17% EE)
3. Mineral supplements		
4. Vitamin supplements		
5. Additives (antibiotics, hormones, probiotics, etc.)		

Nutritive value of various feed ingredients on DM basis

Name of the feedstuff	CP%	EE%	TDN%
Roughages green			
Maize fodder (green)	10	1.79	60-65
Berseem fodder	15.80	1.40	65
Lucerne fodder	23.90	4.30	65
Oats (flowering stage)	10.74	2.50	67
Sorghum fodder	8.23	2.09	56
Napier grass	10.50	6.40	55
Sugarcane fodder	2.65	1.35	35
Cowpea fodder	16.50	3.00	62.1
Sudan grass	9.50	1.92	44.1
Doob/Bermuda grass	4.90	1.20	55
Pearl millet fodder	8.52	1.76	58
Concentrates			
Grains and Seeds			
	CP%	**EE%**	**TDN%**
Barley grain	8.92	2.11	78.0

Name of the feedstuff	CP%	EE%	TDN%
Wheat grain	10.40	2.60	80.0
Maize grain	9.35	3.50	80.0
Gram	20.00	4.10	87.2
Oats grain	9.90	5.40	77.0
Pigeon pea	20.30	1.90	74.1
Pearl millet grain	11.70	5.50	60.6
Broken rice grain	8.30	0.90	78.0
Sorghum grain	11.36	2.17	60.0
Cotton seed	18.10	18.20	---
Flaxseed	19.20	36.10	---
Mustard seed	21.80	37.30	---
Cakes and meals			
Soyabean cake	46.50	6.70	75.0
Soybean meal	49.44	0.40	79.0
Groundnut cake	37.21	8.41	75.0
Groundnut meal	44.44	2.15	70.0
Cottonseed cake	33.2	6.67	69.0
Linseed cake	29.90	4.10	75.0
Mustard cake	33.80	8.50	70
Sunflower seed cake	26.75	6.37	68
Cereal grain milling by-products			
Gram husk	10.15	1.56	55
Wheat bran	13.38	3.39	65.0
Rice bran	14.0	9.80	67.0
Hay			
Oat	7.50	1.47	52.0
Berseem	15.75	1.18	60.0
Sorghum	6.28	1.38	51.4
Soyabean	11.17	5.38	54.0
Silage			
Maize	7.9	-	61.1
Sorghum	5.9	-	51.1
Straw and Stover			
Wheat	3.2	-	48.4
Oat	3.4	-	46.0
Barley	2.2	-	46.9
Rice paddy	3.4	-	46.0

Name of the feedstuff	CP%	EE%	TDN%
Maize stover	3.6	-	45.0
Sorghum stover	3.8	-	56.4
Pearl millet stover	3.5	-	53.4

CHAPTER 24

CALCULATION OF NUTRIENT REQUIREMENTS OF CATTLE, BUFFALO, SHEEP AND GOAT

The success of livestock rearing system depends on the adequacy of the diet to the nutritional needs of animals. Nutrients required for maintenance, growth, milk production, pregnancy, etc. Nutrient requirements of animals are calculated by following feeding standard guidelines. Feeding standards are statements or quantitative descriptions of the amounts of one or more such nutrients needed by animals. Feedings standards are the tables, which indicate the quantities of nutrients to be fed to the various classes of livestock for different physiological functions like growth, maintenance, lactation, egg production and wool growth. The nutrient requirements are generally expressed in quantities of nutrients required per day or as a percentage of diet. For dairy animals, nutrient requirements are generally expressed as separate body functions but in case of poultry and pigs, combined requirements of maintenance and other body functions are given. There are two terms, which has been used, in the feeding standards. One is the nutrient allowance and another is the nutrient requirement.

The various feeding standards of the world which are available for feeding of different categories of livestock are given below:

Feeding standard (country)	Basis of expression	
	Protein	Energy
NRC (USA)	CP,DCP	TDN,DE,NE
ARC (UK)	DCP,AP	DE,ME
Scandinavia	DTP	Feed unit
German	DCP	SE
ICAR (India)	RDP/RUP, MP, CP	TDN, ME

CP= Crude Protein, DCP = Digestible Crude Protein, DTP = Digestible True Protein, AP = Available Protein, TDN = Total Digestible Nutrient, SE = Starch Equivalent, DE = Digestible Energy, ME = Metabolizable Energy, NE = Net Energy.

A. Nutrients requirement of cattle as per NRC guidelines

1. Nutrients requirement of growing females cattle (ADG @ 500 g/day) (adopted from NRC, 2001; Table 14-12)

Bodyweight (kg)	Nutrient requirement				
	CP (g/d)	TDN (kg/d)	ME (Mcal/d)	Ca (g/d)	P (g/d)
100	465.00	1.88	6.70	21	10
150	528.90	2.49	9.10	22	11
200	601.80	3.10	11.30	23	12
250	677.10	3.70	13.40	24	13
300	749.00	4.25	15.30	26	14

2. Nutrients requirement of growing male cattle and mature breeding bulls (adopted from NRC, 1989)

Body weight (kg)	CP (g/d)	TDN (kg/d)	ME (Mcal/d)
Growing male cattle (ADG @ 500 g/day)			
100	392	1.72	6.54
150	525	2.25	8.55
200	573	2.76	10.45
250	598	3.27	12.31
300	707	3.77	14.15
350	823	4.29	16.01
For maintenance of mature breeding bulls			
500	789	4.34	15.79
600	905	4.98	18.10
700	1016	5.59	20.32

3. Nutrients required for working bullocks and bull in service (per head per day) (adopted from NRC, 1989)

Body weight (kg)	Normal work			Heavy work		
	DCP (kg/d)	TDN (kg/d)	ME (Mcal/d)	DCP (kg/d)	TDN (kg/d)	ME (Mcal)
Working bullocks						
300	0.33	3.1	11.2	0.42	4.0	14.4
400	0.45	4.0	14.4	0.52	4.8	17.2
500	0.56	4.9	17.6	0.71	6.4	23.1
Bull in service						
Body weight (kg)	**DCP (kg/d)**	**TDN (kg/d)**	**ME (Mcal/d)**	**Ca(g)**	**P(g)**	**Vitamin A (I.U.)**
500	0.43	4.5	16.2	12	12	21200
600	0.48	5.1	18.2	14	14	25400
700	0.54	5.7	20.5	15	15	29600
800	0.60	6.3	22.5	18	18	33800

4. Nutrients required for mature cattle (per head per day) (adopted from NRC, 1989)

Body weight (kg)	Nutrient requirement		
	CP (g/d)	TDN (kg/d)	ME (Mcal/d)
Maintenance			
400	318	3.13	12.01
450	341	3.42	13.12
500	364	3.70	14.20
550	386	3.97	15.25
Milk production (per kg milk)			
Milk fat%	**CP (g/d)**	**TDN (kg/d)**	**ME (Mcal/d)**
3	78	0.280	1.07
3.5	84	0.301	1.15
4	90	0.322	1.24
4.5	96	0.343	1.32
5	101	0.364	1.40
5.5	107	0.385	1.48
Maintenance during last months of pregnancy			
Bodyweight (kg)	**CP (g/d)**	**TDN (kg/d)**	**ME (Mcal/d)**
400	890	4.15	15.26
450	973	4.53	16.66
500	1053	4.90	18.04
550	1131	5.27	19.37

If the cow is in first and second lactation then nutrient requirement should be increased by 20% and 10% respectively.

B. Nutrients requirements of cattle as per modified Morrison's feeding standard

Body weight (kg)	Nutrient requirement					
	DCP (kg)	TDN (kg)	ME (Kcal)	Ca (g)	P (g)	Vitamin A (I.U.)
Calf growing @ 500 g/d						
45	0.17	0.9	3290	7		2000
70	0.22	1.3	4680	12	10	3000
100	0.28	1.9	6900	13	10	4000
150	0.35	2.6	9360	13	12	6500
200	0.40	3.0	11500	13	12	8500
300	0.47	4.0	12600	13	12	12500
450	0.48	5.0	13600	12	12	17000
Maintenance requirement						
Body weight (kg)	**DCP (kg)**	**TDN (kg)**	**ME (Mcal)**	**Ca (g)**	**P (g)**	**Carotene (mg)**
250	0.168	2.02	7.27	6	6	27
300	0.197	2.36	8.50	7	7	32
350	0.254	2.70	9.72	8	8	37
400	0.277	3.03	10.01	9	9	42
450	0.282	3.37	12.13	10	10	47
500	0.296	3.64	13.28	11	11	52
550	0.336	4.00	14.40	12	12	57
Requirement for the production of 1 kg of milk (to be added to requirement for maintenance and also for growth, if any)						
Milk fat (%)	**DCP (kg/kg milk)**	**TDN (kg/ kg milk)**	**ME (Mcal/ Kg milk)**	**Ca (g/ kg milk)**	**P (g/kg milk)**	
3.0	0.040	0.27	1.97	2.0	1.4	
3.5	0.042	0.29	1.40	2.0	1.4	
4.0	0.045	0.32	1.15	2.0	1.4	
4.5	0.048	0.34	1.22	2.0	1.4	
5.5	0.051	0.36	1.30	2.0	1.4	
6.0	0.057	0.41	1.41	2.0	1.4	
7.5	0.063	0.46	1.66	2.0	1.4	

C. Nutrient requirements of buffaloes

Body weight(kg)	Nutrient requirement	
	CP (g/d)	TDN (kg/d)
Growing buffalo (@500 g/d		
100	373	2.47
150	486	2.86
200	543	3.22
250	604	3.55
Heifers (last 3 months of pregnancy)		
300	538	3.9
350	592	4.2
400	647	4.5
Adult (last 3 months of pregnancy)		
400	644	4.2
450	720	4.5
500	776	4.8
550	832	5.0
600	889	5.5
Adult mature (maintenance)		
350	423	2.8
400	469	3.1
450	512	3.4
500	553	3.6
550	597	3.9
For milk production (for 1 kg milk)		
Fat %	**CP (g)**	**TDN (kg)**
5	98	0.38
6	108	0.43
7	118	0.48
8	128	0.53

D. Nutrient requirements of sheep

1. Nutrient requirements of sheep as per NRC guidelines (NRC, 1984)

Body weight (kg)	Nutrient requirement		
	DM (as % BW)	CP (g/d)	TDN (g/d)
Early weaned lamb (growing @ 100 g/ d)			
5	1.9-2.3	46	140

Body weight (kg)	Nutrient requirement		
	DM (as % BW)	CP (g/d)	TDN (g/d)
10	1.9-2.3	56	230
15	1.9-2.3	76	310
20	1.9-2.3	95	390
Maintenance, growth, non- lactating or pregnant (first 15 weeks of gestation)			
40	2.5-3.0	75	410
50	2.5-3.0	89	480
60	2.5-3.0	102	550
Last weeks of gestation or last 8 weeks of lactation			
20	3.0-4.5	103	640
30	3.0-4.5	148	940
40	3.0-4.5	174	1080
50	3.0-4.5	191	1160
First 8 weeks of lactation			
20	3.3-4.5	105	550
30	3.3-4.5	143	750
40	3.3-4.5	176	930
Maintenance and growth of ram			
30	2.9-3.5	113	720
40	2.9-3.5	137	850
50	2.9-3.5	159	960

2. Nutrients requirements of lambs (ICAR, 1985)

Body weight (kg)	Rate of gain (g/d)	Nutrient requirement				
		DM (g/d)	DCP (g/d)	TDN (g/d)	Ca (g/d)	P (g/d)
Growing lamb						
10	50	400	35	220	2.0	1.5
	100	450	45	280	2.5	1.5
	150	500	55	340	3.0	2.0
15	50	500	45	300	2.8	2.0
	100	600	55	360	3.5	2.5
	150	700	65	450	4.5	3.0
20	50	700	50	400	3.5	2.5
	100	800	70	520	4.5	3.0
	150	1000	80	640	5.5	3.6

Body weight (kg)	Rate of gain (g/d)	Nutrient requirement				
		DM (g/d)	DCP (g/d)	TDN (g/d)	Ca (g/d)	P (g/d)
25	50	800	65	500	4.0	3.0
	100	1100	85	700	5.0	3.5
	150	1200	100	800	6.0	4.0

Maintenance requirements for adult sheep

Body weight (kg)	DM (g)	DCP (g)	TDN (g)	Ca (g)	P (g)
20	500	25	240	3.0	1.8
25	625	30	300	3.2	2.0
30	750	32	360	3.5	2.5
35	875	35	420	4.0	2.8
40	1000	40	480	4.5	3.0
45	1125	42	500	5.0	3.5
50	1250	45	540	5.5	3.8
55	1375	50	600	6.0	4.0
60	1450	54	640	6.5	4.5

Notes: *Breeding rams and pregnant sheep (during last 6 weeks of pregnancy) in reproductive phase requires 50% more nutrients than maintenance. Lactating sheep requires twice the nutrients than that for maintenance during first 6 weeks of lactation and 50% above the maintenance need in the remaining period of lactation.*

E. Nutrient requirements of goats

1. Nutrient requirements of adult goats (NRC, 1981)

Body weight (kg)	Nutrient requirement	
	CP (g/d)	TDN (kg/d)
Maintenance (stable feeding condition, minimal activity, early pregnancy)		
10	22	159
20	38	267
30	51	362
40	63	448
50	75	530
60	86	608

Additional requirement; for different physiological stages		
Physiological stage	**CP (g/d)**	**TDN (kg/d)**
Late pregnancy	82	397
Growing @ 50 g/day	14	100
Growing @ 100 g/day	28	200
Growing @ 150 g/day	42	300
Requirement for milk production		
Fat (%)	**CP (g/d)**	**TDN (kg/d)**
2.5	59	333
3.0	64	337
3.5	68	342
4.0	72	346
4.5	77	351
5.0	82	356

DM requirements

- Maintenance (non producing) = 2.0-2.4% of body weight
- Growing @ 100g/day = 3.2 % of body weight
- Late pregnancy = 3.0- 4.2%
- Lactating = 3.2- 4.5 % of body weight

2. Nutrient requirements of kids growing @ 50.0 g/d (NRC, 1981)

Body weight (kg)	Nutrient requirement				
	DM (g/d)	DCP (g/d)	TDN (g/d)	Ca (g/d)	P (g/d)
10	425	25	275	1.8	1.1
15	600	30	350	2.0	1.2
20	700	35	400	2.4	1.4
25	800	40	450	2.6	1.5
30	950	45	500	2.9	1.6
Maintenance requirements of goats (growth rate 50.0 g/d)					
30	900	35	400	2.0	1.0
40	1200	40	500	2.5	1.2
50	1400	45	560	2.7	1.4
60	1600	55	670	3.0	1.6
70	1800	60	760	3.5	1.8

Specifications for compounded cattle (as per BIS 2023, IS 2052: 2023)

S. No.	Characteristics	Type I	Type II
1	Moisture, percent by mass, *Max*	11	11
2	Crude protein (N × 6.25), percent by mass, *Min*	22	20
3	Crude fat, percent by mass, *Min*	4.0	3.0
4	Crude fibre, percent by mass, *Max*	10	12
5	Acid insoluble ash, percent by mass, *Max*	2.5	3.0
6	Salt (as NaCl), percent by mass, *Max*	1.0	1.0
7	Calcium (as Ca), percent by mass, *Min*	0.8	0.8
8	Total phosphorus, percent by mass, *Min*	0.5	0.5
9	Available phosphorus, percent by mass, *Min*	0.25	0.25
10	Urea, percent by mass, *Max*	1.0	1.0
11	Vitamin A, IU/kg, *Min*	7000	7000
12	Vitamin D3, IU/kg, *Min*	1200	1200
13	Vitamin E, IU/kg, *Min*	30	30
14	Aflatoxin B1, µg/kg, *Max*	20	20
15	Cadmium, mg/kg, *Max*	0.5	0.5

Nutrient requirements for cattle and buffalo (as per ICAR, 2013)

Body weight (kg)	CP (g/day)	TDN (kg/day	Ca (g/day)	P (g/day)	ME (Mcal/day)
150	270	2.10	13.5	9.45	8.40
200	340	2.80	18.0	12.60	11.20
250	420	3.50	22.5	15.75	14.00
300	500	4.20	27.0	18.90	16.80
350	580	4.90	31.5	22.05	19.60
400	660	5.60	36.0	25.20	22.40
450	740	6.30	40.5	28.35	25.20
500	820	7.00	45.0	31.50	28.00
550	900	7.70	49.5	34.65	30.80
600	980	8.40	54.0	37.80	33.60

Lactating buffaloes

Body weight (kg)	CP (g/day)	TDN (kg/day)	Ca (g/day)	P (g/day)	ME (Mcal/day)
400	1200	6.20	48.0	34.0	24.50
450	1320	6.80	54.0	38.0	26.70
500	1440	7.40	60.0	42.0	28.80

Body weight (kg)	CP (g/day)	TDN (kg/day)	Ca (g/day)	P (g/day)	ME (Mcal/day)
550	1560	8.00	66.0	46.0	30.90
600	1680	8.60	72.0	50.0	33.00
650	1800	9.20	78.0	54.0	35.10
700	1920	9.80	84.0	58.0	37.20

Nutrient requirements for Growing Cattle (NRC, 2001)

Growing Cattle

Body weight (kg)	CP (g/day)	NEm (Mcal/day)	NEg (Mcal/day)	Ca (g/day)	P (g/day)	TDN (kg/day)
150	410	3.00	1.70	23.5	13.0	2.10
200	510	3.90	2.30	28.5	15.5	2.80
250	610	4.70	2.80	33.5	18.0	3.40
300	710	5.50	3.40	38.5	20.5	4.10
350	810	6.30	3.90	43.5	23.0	4.70
400	910	7.10	4.50	48.5	25.5	5.30
450	1010	7.80	5.00	53.5	28.0	6.00
500	1110	8.60	5.60	58.5	30.5	6.60
550	1210	9.40	6.10	63.5	33.0	7.20
600	1310	10.20	6.70	68.5	35.5	7.90

Lactating Dairy Cows

Body Weight (kg)	CP (g/day)	NEL (Mcal/day)	Ca (g/day)	P (g/day)	TDN (kg/day)
400	1250	15.50	62.0	35.0	6.20
450	1400	17.50	69.5	39.5	7.00
500	1550	19.50	77.0	44.0	7.70
550	1700	21.50	84.5	48.5	8.50
600	1850	23.50	92.0	53.0	9.20
650	2000	25.50	99.5	57.5	10.00
700	2150	27.50	107.0	62.0	10.70

CHAPTER 25

FORMULATION OF CONCENTRATE MIXTURE

A balance concentrate mixture can be formulated by use of the following methods:

a. Thumb rule method
b. Pearson's square method
c. Algebraic method
d. Linear programming

Thumb rule method

Generally as per thumb rule, the concentrate mixture can be prepared by mixing one-third of each of cereals, oil cakes and cereal or pulse byproducts. Broadly according to this method, the concentrate mixture may also be prepared as per the following flexible formula.

Ingredients	Parts (kg)
Oil cakes	20-35
Cereals/millets	25-60
Cereal milling by-products (brans and chuni)	10-35
By-products (molasses)	5-20
Mineral mixture	2
Common salt	1
Vitamin A and D	10-15 g

The above concentrate mixture will contain 15% DCP and 70% TDN.

Pearson's square method

This method may be applied under the following conditions:

1. When only two feeds are involved
2. When more than two feeds are involved.
3. When a definite percentage of a particular ingredient is involved.

1. When only two ingredients/feeds are involved:

To know that what combination of two feeds in a mixture will supply the specific amount of a particular nutrient, the following steps will be followed:

1. Draw a square on left side of the page.
2. Insert the percentage of nutrient (crude protein) desired in the middle of the square.
3. Place the basal feed with its CP% on the upper left corner and supplement feed with CP% on the lower left corner of the square.
4. Subtract the CP% in the basal feed. From CP% desired in the mixture and place the difference on the corner of the square diagonally opposite to the basal feed. This is the amount of supplement in the mixture.
5. Subtract the CP% desired in the mixture from the CP% in the supplement and place the difference on the corner of the square diagonally opposite to the supplement. This is the amount of basal feed in the mixture.
6. The above remainders represent the amount of two feeds in a mixture weighing equal to the sum of two remainders having a desired percentage of CP. The amounts are then converted into percent to simplify mixing.

Exercise 1. A farmer wants to prepare a concentrate mixture containing 20% CP by the use of maize as basal feed (8.5% CP) and GNC as protein supplement (40% CP) for feeding a group of animals. Calculate the proportion of these ingredients in mixture?

Solution: As per the above steps:

Basal feed | 40-20=20.0 (Basal feed)

Maize 8.5

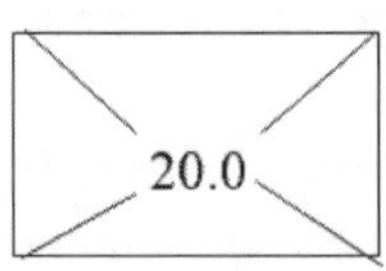

GNC 40 | 20-8.5=11.5 (Protein supplement)

Total = 31.5

$$\text{Amount of basal feed (maize)} = \frac{20}{31.5} \times 100 = 63.49 \text{ kg}$$

$$\text{Amount of protein supplement (GNC)} = \frac{11}{31.5} \times 100 = 36.51 \text{ kg}$$

Ans. Maize 63.49 parts or kg and GNC 36.51 parts or kg in 100.00 kg mixture.

2. When more than two ingredients/feeds are involved:

Exercise 2. A dairy man desires to use a mixture of barley (10.5% CP), oats (12.0% CP) and groundnut cake (42% CP) in formulating a mixture with 18% CP for his animals using barley and oats in the ratio of 2:1. Find out the quantities of each of the ingredient?

Solution:

1. Draw a square on left side of the page.
2. Put the % CP desired in the middle of the square.
3. Separate the feeds into two groups. Specify the proportion of each feed in each group and calculate the average CP% in each group. The weighed average CP in the barley and oat mixture in the ratio of 2:1 would be

   2 barley: 2 × 10.5 = 21.0, 1 oats: 1 × 12.0 = 12.0, total = 33.0

   Weighed average CP (%) = 33/3 = 11.0
4. Place "2 barley and 1 oats" with average % CP (11.0) on the upper left corner of the square and supplement with its % CP (42) on the lower left corner.
5. Subtract diagonally and record the differences as in the previous example
6. Subdivide the final composite figure for barley and oats by multiplying with 2/3rd and 1/3rd, respectively.

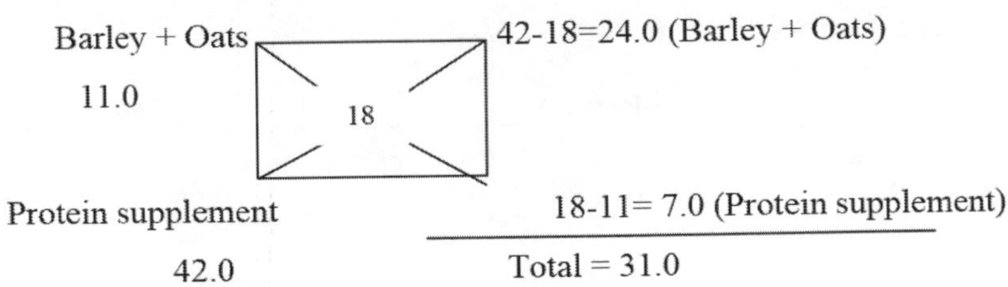

$$\text{Amount of barley + oats} = \frac{24}{31} \times 100 = 77.42 \text{ kg}$$

$$\text{Amount of barley + oats} = \frac{7}{31} \times 100 = 22.58 \text{ kg}$$

Out of 77.42 parts of barley + oats, the amount of barley and oats separately will be

Barley = 77.42 × 2/3 = 51.61 kg

Oats = 77.42 × 1/3 = 25.81 kg

Ans. The final mixture will have barley 51.61 kg, oats 25.81 kg and groundnut cake 22.58 kg.

4. When a definite percentage of a particular ingredient is involved

Exercise 3. A farmer desires to formulate a 20% CP mixture for a cow using maize (8.5% CP), oats (12.0% CP) and soybean meal (45.8% CP) and a mineral and vitamin supplement. He wants to include in the mixture exactly 20% oats and 3% mineral, salt and vitamin supplement?

Solution

Under such circumstances, the percentage of maize and soybean oil meal in the remaining 77 parts of the mixture with overall CP of 20% can be calculated as follows:

Since 20 kg of each 100 kg of the mixture is oats so in 100 kg it will supply 2.4 kg CP. The mineral and vitamin supplement will not contribute towards CP. Hence 20 kg oats plus 3 kg mineral, salt and vitamin supplement will supply 2.4 kg protein. So the remainder 17.6 kg (20-2.4) CP should come from 77 kg of maize and soybean oil meal. The amount of maize and soybean oil meal in 77 kg can be determined as follows:

Calculate % CP needed in the maize and soybean oil meal combination to provide 17.6 kg of CP per 77 kg as follows:

$$= \frac{17.6}{77} \times 100 = 22.86\%$$

Then proceed as in the first case

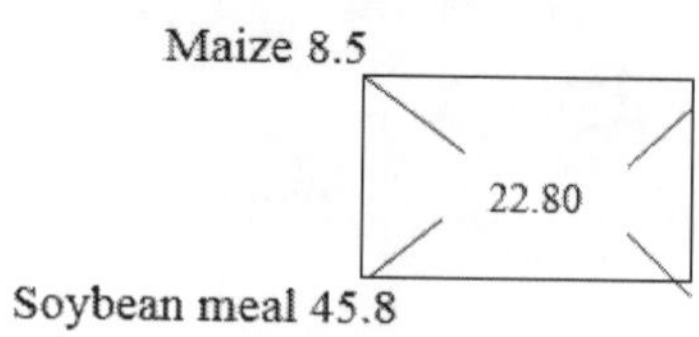

45.8-22.86=22.94 (Maize)

22.86-8.5 = 14.36 (Soybean oil meal)

Total = 37.30

On per 77 parts basis

$$\text{Amount of maize} = \frac{22.94}{37.30} \times 77 = 47.36 \text{ kg}$$

$$\text{Amount of soybean meal} = \frac{14.36}{37.30} \times 77 = 29.64 \text{ kg}$$

Ans. So, 100 kg of the final mixture will have maize 47.36 g, soybean meal 9.64 kg, oats 20 kg and mineral, salt and vitamin supplement 3.0 kg.

Methods for computation of concentrate mixture

1. Single Pearson's square method

This method is used for compounding of concentrate mixture from two feed ingredients only. In this method either protein or energy density in the compound feed is calculated.

Exercise 4. Calculation for the fixing of the protein percentage in the concentrate mixture by taking two ingredients?

Solution

For this purpose protein content of one ingredient should be higher and that of the other ingredient lower than the proposed protein content of the concentrate mixture. For the compounding of feed mixture of 20% CP on DM basis, the feed ingredients available are ground nut cake (GNC) containing 45% CP and 75% TDN and maize grain containing 10% CP and 85% TDN on DM basis. Now, place the values of CP% in the two ingredients at point A and B of the square and desired level of CP% in the centre. After this, subtract the value of the ingredients and the centre value (lower value will always be subtracted from the higher value diagonally). Record the difference without any symbol at point C & D, which is total part of the compounded feed and the values at C is the part of GNC and that at D part of maize grain as shown in the Pearson's square.

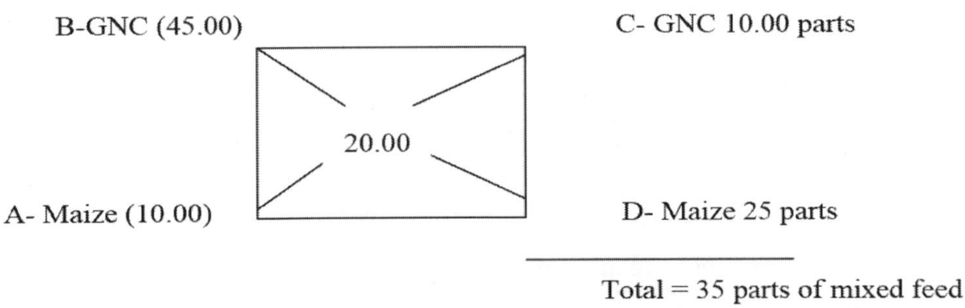

Since in 35 parts mixed feed GNC is 10 parts, Hence in 100 parts GNC will be: 10/35 × 100= 28.57%, The percentage of maize grain will be: 100 - 28.57= 71.43%. Thus, the feed mixture will be prepared by mixing 28.57% GNC with 71.43% maize grain and the mixture will contains 20% CP on DM basis. Now, the calculation of TDN in the mixture: TDN %= 28.57 × 75/100 + 71.43 × 85/100 = 21.42 + 60.72 = 82.14%.

2. Algebraic method or simultaneous equation for computation of concentrate mixture

Simple equation like A + B = 100 or A + B + C = 100 can be used for calculation of concentrate mixture. The equation of two ingredients are used for determining the level of any one nutrient i.e. either CP or TDN/DE/ME, whereas, that with three ingredients may be used for the calculation of both protein (CP/DCP) and energy (TDN/DE/ME) contents in the mixture, provided the ingredients selected for the preparation of concentrate mixture contain lower as well as higher values of the nutrients to be worked out.

Exercise 5. Compute a concentrate mixture of 20% CP content with help of simultaneous equation from two ingredients i.e. maize grain and GNC. Also determine the TDN %?

Name of the ingredients	CP %	TDN%
Maize	10.00	85.00
GNC	45.00	75.00

Solution

Suppose maize is A and GNC is B feed.

Then, A + B = 100 ————(i)

0.1 A + 0.45 B = 20 ————(ii)

Now multiplying equation (i) with by o.1 to get the following equation,

A = 0.1 B = 10 ————(iii)

Now subtract equation (iii) from equation (ii)

A + 0.45 B =20

A + 0.10 B = 10

0.35 B = 10

B = 10/0.35= 28.57 %

Hence A = 100 – 28.57 = 71.43

Ans. Thus the concentrate mixture will be prepared by mixing 71.43 kg of maize grain and 28.57 kg of GNC. TDN % in the concentrate mixture will be:

TDN % = 28.57 x 75/100 + 71.43x 85/100 = 82.14%

Exercise 6. Prepare a concentrate mixture with desired CP and TDN % by double Pearson's square method?

a. Select the ingredients in such a way that putting the values in the squares should give desired CP % and more than desired TDN % for mixture-1. Calculate the TDN % in the mixture.

b. Select the ingredients in such a way that putting the values in the squares should give desired CP % and less than desired TDN % for mixture-2. Calculate the TDN % in the mixture.

c. Put the values of TDN percentage in the mixture-3 of mixture-1 and mixture-2 so that the mixture-3 should give desired CP% and TDN% and calculate the parts (%) of 1st and 2nd mixture to be used in 3rd mixture.

Calculate the ingredients used in 3rd mixture of 1st mixture under:

$$\frac{\text{\% ingredients used in 1st mixture x \% of 1st mixture used in 3rd mixture}}{100}$$

d. Similarly, calculate ingredients of 2nd mixture used in 3rd mixture.

e. Total the ingredients used in mixture-3rd i.e. having desired CP% and TDN%.

f. Check the CP and TDN % in the concentrate mixture by multiplying the quantity of ingredients with CP and TDN% in the ingredients.

Exercise 7. Prepare a concentrate mixture having 12% CP and 74% TDN by using double Pearson square method with the available ingredients?

Ingredients	CP %	TDN %
Corn	10	80
Cotton seed meal	40	68
Alfalfa meal	15	55

Solution

a) Concentrate mixture- 1 having 12 % CP and higher than desired TDN%

Ingredients	CP %	Parts (%)	TDN %
Corn	10	93.33	93.33 x 0.80 =74.64
CSM	40	6.67	6.67 x 0.68 =4.45
Total		100.00	79.18

b) Concentrate mixture- 2 having 12 % CP and lower than desired TDN%

Ingredients	CP %	Parts (%)	TDN %
Corn	10	60	60 x 0.80 =48.00
alfalfa	15	40	40 x 0.55 = 22.00
Total		100.00	70.00

c) Concentrate mixture- 3 having 12 % CP and 74% TDN

Mixture	TDN %	Parts (%)
1	79.00	43.48
2	70.00	56.42
Total		100.00

d) Ingredients required in concentrate mixture having 12% CP and 74% TDN

1. Corn in concentrate mixture- 1 = 93.33%
2. Corn in concentrate mixture- 3 = 93.33 x 43.48/100= 40.58%
3. Cotton seed meal in concentrate mixture-1 = 6.67%
4. Cotton seed meal in concentrate mixture-1 = 2.90%
5. Corn in concentrate mixture -2 = 60.00%
6. Corn in concentrate mixture -2 = 22.61%
7. Alfalfa in concentrate mixture-2 = 40.00%
8. Alfalfa in concentrate mixture-3 = 22.61%
9. Total ingredients used in concentrate mixture-3
10. Corn = 40.58 + 3.91 = 74.00
11. Cotton seed meal = 2.90
12. Alfalfa =22.61

e) **TDN and CP percentage in 3rd mixture**

Ingredients	CP %	TDN %
Corn	74.49x10/100=7.449	74.49x80/100=59.592
alfalfa	1.160	1.972
Total	3.391	12.436
Total	12.00	74.00

Examples of concentrate mixtures

Concentrate mixture	Ingredients	Parts or %	DCP (%) and TDN (%)
1	Mustard oil cake	30 %	16% DCP 80% TDN
	Barley	37 %	
	Wheat bran	30 %	
	Mineral mixture	2 %	
	Common salt	1 %	
2	Groundnut cake	20 %	14.8% DCP 70.9% TDN
	Wheat bran	30 %	
	Barley	32 %	
	Oats	15 %	
	Mineral mixture	2 %	
	Common salt	1 %	

Concentrate mixture	Ingredients	Parts or %	DCP (%) and TDN (%)
3	Wheat bran/rice bran	35%	15% DCP 72.0% TDN
	Dal chuni (moong, gram, urd, arhar etc.)	17 %	
	Barley	45 %	
	Mineral mixture	2 %	
	Common salt	1 %	
4	Groundnut cake	17 %	15.6% DCP 72.0% TDN
	Wheat bran/rice bran	30 %	
	Maize/barley/oats	30 %	
	Cotton seed cake (un-decorticated)	20 %	
	Mineral mixture	2 %	
	Common salt	1 %	

CHAPTER 26

COMPUTATION OF RATION FOR RUMINANTS AT DIFFERENT BODY FUNCTIONS

Nutrients requirement of animals depends upon age, breed, sex, stage of production, level of production, stress conditions, etc. Nutrients requirement of farm animals generally considered for maintenance, growth, pregnancy, milk production and some time work and exercise.

A. Computation of ration for maintenance

Exercise 1: Compute a ration for a dry, non-pregnant cow weighing 400 kg:

(a) With maintenance quality non- leguminous fodder (Maize)

1. The nutritive value of maize fodder is CP-10%, TDN-65%
2. Nutrients requirements of a dry, non-pregnant cow weighing 400 kg are DM @1.5-2.0 % of body weight, CP 0.318 kg/d, and TDN 3.13 kg/d.

So feeding schedule will be prepared as below

Feedstuffs	DM (kg/d)	As fed (kg/d)	CP (kg/d)	TDN (kg/d)
Maize Fodder	4.82	19.28	0.482	3.13
Requirement	6-8		0.318	3.13

Note: *Though the protein and energy requirements are fulfilled, animal is still deficit in DM. Thus 5-10 kg extra fodder should be provided to the animal to satisfy its hunger.*

(b) With leguminous fodder and straw

The nutritive value of the available feeds is given below:

Feedstuffs	CP (%)	TDN (%)
Berseem	13.00	66.00
Wheat Straw	3.20	48.40

Amount of berseem (DM) required satisfying the protein requirement.

$$= \frac{100}{13} \times 0.318 = 2.45 \text{ kg}$$

TDN supplied from 2.45 kg Berseem = 1.47 kg

Balance TDN to be provided from wheat straw = 3.13 –1.47 =1.66 kg

Amount of wheat straw (DM) required $4.82 \times \frac{100}{148.4} \times 1.66 = 3.43 \text{ kg}$

So feeding schedule will be:

Feedstuffs	DM (kg)	Fresh feed (kg)	CP (kg)	TDN (kg)
Berseem	2.45	9.80	0.318	1.47
Wheat straw	3.43	3.81	0.109	1.66
Total	5.88	13.61	0.427	3.13
Requirement	6 to 8		0.318	3.13

(c) With concentrate mixture + straw

Under such situation, concentrate mixture feeding is necessary to meet out the protein requirement of the animal nutritive value of the available feedstuffs is given below:

Feedstuffs	CP (%)	TDN (%)
Wheat Straw	3.2	48.4
Concentrate	20.0	70.0

Solution:

Step-1: Amount of concentrate mixture required to meet out the CP requirement

$$= \frac{100}{20} \text{X } 0.318 = 1.59 \text{ or } 1.6 \text{ kg}$$

Step-2: TDN supplied by 1.6 kg concentrate mixture

$$= \frac{70}{100} \text{ X } 1.6 = 1.12 \text{ kg}$$

Balance TDN (3.13 - 1.12 = 2.01 kg) is to be supplied through wheat straw

Step-3: Amount of wheat straw required to supply 2.01 kg TDN

$$= \frac{100}{48.4} \text{X } 2.01 = 4.15 \text{ kg}$$

Step-4: Therefore, feeding schedule would be as below

Feedstuffs	DM (kg)	Fresh feed (kg)	CP (kg)	TDN (kg)
Concentrate mixture	1.60	1.80	0.32	1.12
Wheat straw	4.15	4.61	0.15	2.01
Total	5.75	6.41	0.47	3.13
Requirement	6-8		0.318	3.13

Note: *In such cases as indicated above some more quantity of straw/fodder is given to meet out the DM requirement.*

B. Computation of ration for gestation

Exercise 1. Compute a ration for a cow weighing 400 kg. The cow is in advanced stage of pregnancy. The available feedstuffs along with their nutritive value are as follows:

Feedstuffs	CP (%)	TDN (%)
Wheat straw	3.20	48.4
Concentrate mixture	20.00	70.0

Solution

Step-1: Calculation of nutrient requirements

a. DM - @ 1.5 to 2% of body weight i.e. 6-8 kg

b. CP and TDN requirements

	CP (kg)	TDN (kg)
Maintenance + gestation	0.890	4.15

Step-2: Calculation of the quantities of feeds and fodders required to meet out the nutrient requirements. Amount of concentrate mixture required to supply the required CP would be

$$= \frac{100}{20} \text{X } 0.890 = 4.45 \text{ kg}$$

TDN supplied by 4.45 kg of concentrate mixture

$$= \frac{70}{100} \text{X } 4.45 = 3.11 \text{ kg}$$

Step-3: Balance TDN to be supplied by wheat straw =4.15 –3.11 =1.04 kg

Amount of wheat straw required to supply 1.04 kg of TDN

$$= \frac{100}{48.4} \text{ X } 1.04 = 2.15 \text{ kg}$$

Step-4: Therefore, feeding schedule would be as below:

Feedstuffs	DM (kg)	Fresh feed (kg)	CP (kg)	TDN (kg)
Wheat straw	2.15	2.39	0.076	3.11
Concentrate mixture	4.45	4.94	0.890	1.04
Total	6.60	7.33	0.966	4.15
Required	6-8		0.890	4.15

C. Computation of ration for growth

Exercise 2. Compute a ration for a calf weighing 200 kg and gaining @ 500 g/d.

Solution: Nutrient requirement of the calf is:

DM @ 2-2.5% of live Wt. = 4-5 kg

CP = 0.573 kg & TDN = 2.76 kg

The available feedstuffs along with their CP and TDN contents are as follows

Feedstuffs	CP (%)	TDN (%)
Concentrate mixture	20	70
Berseem	13	60
Wheat straw	3.20	48.4

Solution

Step-1: Requirements: same as above.

Step-2: Calculation of quantities of feeds and fodders needed to supply the required nutrients.

Distribution of DM: amount of total DM required = 4.5 kg

For growing animals, depending upon growth rate, higher amount of concentrate (say 40-50% of total DM) should be given. The amount of concentrate will further depend upon the type of green fodder available.

In the present case, due to the availability of green berseem the concentrate has been given as 40% of total required.

Therefore, amount of conc. provided = 0.40 x 4.5 = 1.8 kg

Nutrients supplied by 1.8 kg concentrate mixture

1.8 kg concentrate mixture =0.360 kg CP and 1.26 kg TDN

Remaining CP would be = 0.573-0.360 kg = 0.213 kg

Amount of berseem required to meet out the remaining CP (0.213 kg) would be = 1.64 kg (approx.)

Nutrients supplied by Berseem:

s1.64 kg berseem = 0.213 CP and 0.984 kg TDN

Balance TDN to be supplied by wheat straw

= 2.76-(1.26+0.984) = 0.516 kg TDN

Amount of wheat straw required

$$= \frac{100}{48.4} \text{ X } 0.516 = 1.07 \text{ kg}$$

Feeding schedule will be as follows:

Feedstuffs	DM (kg)	Fresh feed (kg)	CP (kg)	TDN (kg)
Concentrate mixture	1.80	2.00	0.360	1.260
Berseem	1.64	6.56	0.213	0.984
Wheat straw	1.07	1.19	0.034	0.516
Total	4.51	9.75	0.607	2.760

D. Computation for milk production

Exercise 3. Compute a ration for a cow weighing 500 kg, yielding 10 liters of milk with 4.0% fat.

Feedstuffs	CP (%)	TDN (%)
Oats	10	67
Concentrate mixture	20	70
Wheat straw	3.2	48.4

The available feedstuffs along with their CP and TDN contents are:

Solution:

Step-1: Calculation of nutrient requirements

DM @ 2.5% of live wt. = 12.5 kg, Requirements of CP and TDN

	CP (kg)	TDN (kg)
For maintenance	0.364	3.70
For milk production	0.900 (0.090x10)	3.22 (0.322x10)
Total	1.264	6.92

Step-2: Calculation of the quantities of different feeds and fodders to supply the nutrients set up in step-1

		2/3rd dry (5.5 kg)
DM	2/3rd roughage (8.5 kg)	1/3rd green (3.0 kg)
	1/3rd concentrate (4.0)	

DM Partitioning

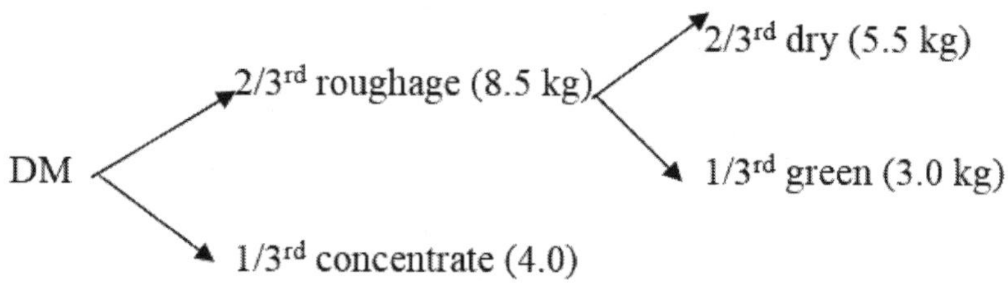

So feeding will be as below:

Feedstuffs	DM (kg)	Fresh feed (kg)	CP (kg)	TDN (kg)
Concentrate mix	4.0	4.9	0.800	2.80
Wheat straw	5.5	6.1	0.176	2.66
Oats	3.0	15.0	0.300	2.01
Total	12.5	26.0	1.276	7.47
Required	12.5		1.264	6.92

In the prepared feeding schedule, CP and TDN is slightly more than required. This kind of margin can be kept as some amount of offered feed may be wasted by the animal during feeding.

E. Computation for work

The nutrient requirements of working animals depend upon the amount of work performed. For heavy work i.e hours of carting and six hours of ploughing in a day, the animal needs higher amount of easily digestible carbohydrates are ration. Out of the total gross energy intake per day, an animal working for a full day yields 9% of the intake as work. Which however, is subject to further variation, depending on type of work, speed of working, fatigue, practice, and the protein requirement though is not increased with the increased workload but to maintain a proper nutritive ratio, amount of dietary protein has to be increased.

Exercise 4. Compute a ration for bullock weighing 400 kg on normal work. The available feedstuffs along nutritive value arc as follow:

Feedstuffs	CP (%)	TDN (%)
Wheat straw	3.2	48.4
Napier grass	5.4	58.0
Concentrate mixture	20	70

Solution:

Step-1: Calculation of nutrient requirements

DM required = @ 2.5% of weight = 10 kg, CP required = 0.56 kg, TDN required = 4.0 kg.

Sept-2: Calculation of the amount of feedstuffs

DM Partitioning

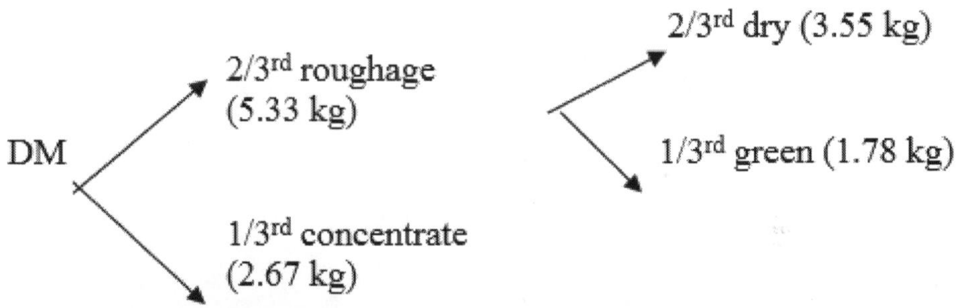

Feeding schedule will be as follows:

Feedstuffs	DM (kg)	Fresh feed (kg)	CP (kg)	TDN (kg)
Wheat straw	3.55	3.94	0.113	1.72
Napier grass	1.78	8.90	0.096	1.03
Concentrate mix.	2.67	2.96	0.534	1.86
Total	8.00	15.80	0.743	4.61
Required	8.00		0.56	4.00

CHAPTER 27

FORMULATION OF RATION DURING SCARCITY PERIOD FOR CATTLE

During the lean or scarcity periods of May and mid July and November–December, limited quantity of green fodder is available. Feedstuffs that can be include in conventional feed can be used for feeding the animal during scarcity period is called as scarcity fodder. The shortage in livestock feed is caused due to flood and draught and there is no single year in our history in which flood has not occurred. In most parts of our country, drought occurs during drought period, shortage of animal feed especially roughages is encountered. This has been an adverse effect on feed resources as dried roughages alone account for over 50-80 % dry matter and TDN requirement for ruminant e.g. sheep, goat, cattle.

Large variety of plants which are not conventional, and not included in feeds for livestock, is used for feeding the cattle and other ruminants during scarcity period. These include large variety of tree, leaves, shrubs, weeds, grasses. Considerable palatable varieties of leaves are conventional and used foe sheep, goat. Certain varieties also fed to cattle and buffaloes. Leaves of different species have different nutritional characteristics. In general, early stage of growth of leaves contain fairly high percentage of crude protein comparatively low fibre when the vegetation advances in age, there is a progressive decrease in protein and increase in CF. The leaves are generally rich in Ca and P is generally wide. Due to presence of tannic acid, the digestibility of protein in the leaves is low. The stem and leaves of number of other plants like shrubs and leaves of grasses were also investigated as food for livestock. Important variety of this type are different type of grasses cultivated and uncultivated.

During periods of feed scarcity in India, such as during droughts or dry seasons, formulating a suitable ration for cattle becomes crucial to ensure their health and productivity. The scarcity typically results in a reduced availability of high-quality forages and feed resources, necessitating adjustments to meet the cattle's nutritional needs effectively. During these times, it is essential to provide adequate crude protein and energy. The crude protein requirement for maintenance is generally around 8-12%, while lactating cattle may need up to 12-14% protein. TDN should be maintained at approximately 55-65% to ensure sufficient energy intake.

Alternative feed resources become vital when traditional feeds are scarce. For example, crop residues like wheat straw, rice straw, and maize stover are commonly used. These materials are typically low in protein and energy but can be enhanced by adding supplements. Urea, at a rate of 1-2% of the total diet, can be added to crop residues to improve their protein content and digestibility. By-products such as rice bran, wheat bran, and oilseed cakes also serve as important sources of protein and energy. These should be used judiciously to balance the ration without causing nutritional imbalances.

A typical ration formulation during scarcity might include 60% crop residues, 20% concentrate feeds like wheat bran, 10% oilseed cakes, and 10% urea-treated straw or silage. This combination helps in providing a balanced diet that meets the basic nutritional requirements of cattle. Additionally, incorporating a mineral mixture is crucial to ensure that cattle receive adequate levels of essential minerals such as calcium and phosphorus, which might be deficient in the available feed resources.

Implementation involves preparing feeds and supplements ahead of time to ensure a steady supply during scarcity. It is also important to manage feeding practices effectively by offering smaller, more frequent meals to improve feed intake and digestion. Regular monitoring of cattle health and feed efficiency helps in making necessary adjustments to the ration and addressing any nutritional deficiencies or health issues promptly.

Few examples of ration for adult non-producing cattle in scarcity period:

1) Example-1

1. Take 2 kg of bagasse
2. Add 400 g of molasses
3. Add 8 kg sugarcane tops
4. 22 g urea
5. 30 g NaCl
6. 50 g mineral mixture.

This preparation is made and fed to animal in scarcity. For an adult animal it is sufficient.

2) Example-2

1. 85 kg baggasse
2. 2 kg urea
3. 12 kg molasses
4. 1 kg mineral mixture

The maximum level of urea is 3 %. Thus 100 kg of feed should contain less than 3 kg of urea.

3) Example-3; 100 kg straw

1. 2.5 kg urea solution
2. 10 kg molasses
3. 2 kg mineral mixture

4) Example-4

1. 4 kg Sugar cane bagasse
2. 0.75 kg molasses
3. 2 kg Dry leaves
4. 30 g urea
5. 20 g mineral mixture

5) Example-5

1. 3 kg dry grass/ moonj / kans/ sugarcane bagasse
2. 6 kg dry leaves
3. 0.5 kg molasses
4. 30 g urea
5. 30 g salt
6. 20 g mineral mixture

6) Example-6

1. 5 kg dry fodder
2. 40-50 g urea
3. 1 kg molasses
4. 30 g salt
5. 30 g mineral mixture

7) Example-7

1. 6 kg sugar cane bagasse
2. 3 kg dry leaves
3. 0.75 kg molasses
4. 20 g urea
5. 10-20 g salt
6. 10-20 g mineral mixture
7. 0.5 g vitamin mixture

CHAPTER 28

VISIT TO CATTLE FEED PLANT

Production Process

The major operations involved in the production of animal feed are: raw materials preparation, primary crushing, assorting and measuring, molasses mixing, fine crushing, pellet making, and packaging. Raw and auxiliary materials are first charged into silos and tanks where they are made ready for further processing. They are then processed by primary crusher. Crushed materials are further separated by means of a sifter, and then stored in the assorting tanks according to the kind of raw materials.

Detailed Production Process of a Cattle Feed Plant

Raw Material Handling and Storage

The production of cattle feed begins with the careful management of raw materials, crucial for ensuring the quality and nutritional value of the feed.

1. Sourcing and Procurement

Raw Materials: The primary ingredients for cattle feed include grains such as corn and barley, oilseeds like soybean meal, and forages such as alfalfa. Each ingredient is selected based on its nutritional profile. Corn contains 8-10% protein, 3-4% fat, and 70-73% starch, making it a high-energy source. Barley offers 12-14% protein, 2-5% fat, and 60-65% starch, which provides a balanced energy and protein source. Soybean meal provides 44-48% protein, 1.5-2.5% fat, and 6-7% fiber, important for protein enrichment etc.

Storage

- **Dry Storage**: Ensuring proper storage conditions is critical to prevent spoilage and maintain feed quality:

- **Temperature Control**: Raw materials are stored at temperatures below 30°C (86°F) to inhibit mold growth and degradation.
- **Humidity Control**: Relative humidity is maintained below 70% to prevent moisture-related issues.
- **Silos**: Storage silos have capacities ranging from 50 to 500 tons, accommodating different scales of operation.
- **Bulk Storage**: For large-scale operations, bulk storage bins can handle up to 1,000 tons of material. These bins are designed to prevent contamination and spoilage, ensuring that feed remains in optimal condition until use.

Handling Equipment: Efficient material handling is essential for smooth operations:

- **Conveyor Systems**: These can move up to 50 tons of material per hour, ensuring a steady flow of raw ingredients.
- **Elevators**: Used to transport materials vertically, with capacities ranging from 20 to 50 tons per hour, these elevators facilitate the efficient movement of materials within the plant.

2. Raw Material Preparation

Cleaning

Pre-cleaning: Before processing, raw materials are cleaned to remove larger contaminants:

- **Capacity**: Pre-cleaning equipment can handle 10-15 tons per hour.
- **Efficiency**: This stage achieves debris removal with an efficiency rate greater than 99%.

Fine Cleaning: Fine cleaning further removes smaller impurities:

- **Air Aspirators**: These devices reduce impurity levels to less than 0.5%.
- **Screens**: Used in conjunction with aspirators, screens help in the fine separation of particles.

3. Grinding

Hammer Mills: Essential for reducing the particle size of raw materials:

- **Particle Size**: Achieves a size range of 0.5-1.0 mm, suitable for many feed formulations.
- **Capacity**: Processes 5-10 tons per hour, depending on the mill's configuration.

Roller Mills: Used to create a finer particle size, especially for specific feed types:

- **Particle Size**: Produces particles of 0.8-1.5 mm, depending on settings.
- **Capacity**: Handles up to 20 tons per hour, allowing for high throughput.

4. Feed Mixing

Ingredient Proportioning

Feed formulation requires precise balancing of ingredients to meet nutritional needs:

- **Protein Content**: Typical cattle feeds contain 12-20% crude protein, critical for growth and milk production.
- **Energy Content**: Feeds are formulated to provide 70-80% total digestible nutrients (TDN), ensuring adequate energy levels for cattle.

Mixing types:

Batch Mixing: This method involves mixing ingredients in batches:

- **Capacity**: Mixers handle 5-15 tons per batch, providing flexibility in formulation.
- **Mixing Time**: Typically 5-10 minutes, ensuring thorough blending.
- **Continuous Mixing**: Suitable for large-scale operations requiring consistent mixing:
- **Capacity**: Can process 30-50 tons per hour, allowing for high-volume production.

5. Processing

Conditioning

Steam Conditioning: Enhances the physical properties of feed and its digestibility:

- **Temperature**: Feed is conditioned with steam at 85-90°C (185-194°F).
- **Conditioning Time**: 30-60 seconds is sufficient to improve binding and digestibility.

6. Pelleting

Pelletizing: Converts mixed feed into a solid form:

- **Temperature**: Pellets are produced at temperatures up to 85°C (185°F).
- **Pellet Diameter**: Typically ranges from 3-8 mm, depending on the feed requirements.
- **Capacity**: Pellet mills can produce 10-20 tons per hour, enabling efficient processing.

7. Cooling: Post-pelleting cooling is crucial to prevent condensation:

- **Temperature Reduction**: From 85°C to 30-35°C.
- **Purpose**: Cooling reduces moisture and enhances pellet durability.

8. Quality Control

Sampling and Testing

Nutritional Analysis: Ensures feed meets nutritional specifications:

- **Protein Content**: Maintained within ±1% of target levels.
- **Fat, Fiber, and Mineral Content**: Tested to meet required levels.

Physical Testing: Evaluates the quality of the finished feed:

- **Moisture Content**: Should be less than 12%, preventing spoilage.
- **Pellet Durability**: Must be above 90% to ensure feed remains intact during handling.

9. Storage of Finished Feed

Finished Feed Storage: Ensures feed remains in good condition until use:

- **Capacity**: Storage bins and silos have capacities ranging from 100 to 500 tons, depending on plant size.

Figure 1. Diagram of feed plant

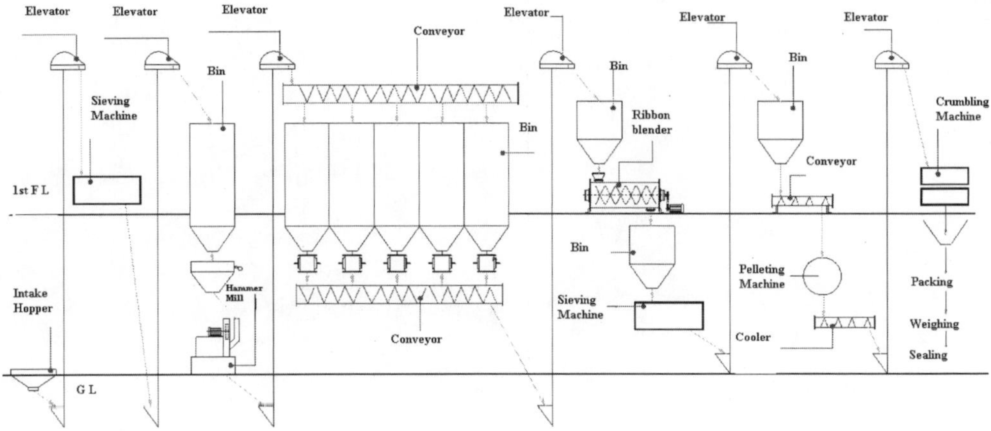

Figure 2. Outlay of feed manufacturing unit

CHAPTER 29

PRINCIPLES OF COMPOUNDING AND MIXING OF FEEDS

"A ration is a feed offered to a given animal during a time period of 24 hours. The feed may be given at a time or in portions at intervals".

Classification of ration:

1. **Balanced ration**: It is that feed mixture which contains all the essential nutrients in right quality and quantity as needed by the animal for maintenance, growth, work and production.

c. **Maintenance ration:** The amount of balanced ration required to fulfill the maintenance need of a particular animal is termed as maintenance ration or maintenance requirement or maintenance allowance.

d. **Production ration:** The amount of feed mixture given to a growing, working or producing animal over and above maintenance requirement is known as production ration.

e. **Ideal ration:** The feed mixture containing all the essential components of a balanced ration which has got ability to fulfill all the needs of a particular animal when given according to his body weight is called as an ideal ration.

Principles of rationing:

1. The ration should be highly digestible and well balanced in terms of energy, protein, minerals and vitamins.
2. The feeds composing the ration should be good and sound and a variety of such feeds should be used to formulate the ration for better balance of nutrients as well as palatability.

3. The ration should be fairly laxative and bulky to stimulate the walls of digestive track for maximum secretion and action of digestive juices.
4. The feed ingredients used in ration formulation should be easily available and cost effective.
5. Practical diet formulation must be sufficiently flexible to accommodate price and feedstuff available while retaining the necessary nutritive balance and adequacy.
6. Ration must be properly prepared. Hard grains like maize, barley, sorghum etc. should be grounded before feeding.
7. The average particle size of diet should be 800 microns for poultry and between 650 to 750 microns for swine for better utilization of feeds.
8. In the ration formulation the crude fibre (CF) contents of diet should be adjusted between 6 to 8 %.
9. Ration must contain all limiting amino acids in balanced amount. Animal protein supplements are rich source for lysine and methionine. Among vegetable protein sources, soybean meal is a rich source for lysine. Further, lysine & methionine are also available in synthetic form.
10. The Na Content usually is not calculated during ration formulation. The addition of 0.5% and 0.6% common salt in pig and poultry ration, respectively meets the requirements.
11. If feed contains antibacterial compound, the withdrawal period of that compound must be kept in mind. This is the period of time that medicated feed must be removed from pig/poultry diets before you slaughter them.
12. Swine and poultry ration usually are formulated using cereal grains as base ingredients because they are low in fibre (CF) and high in energy.
13. The ration should be formulated and fed according to the requirement of the animals. Feeding less or in excess, both are detrimental to health.
14. In order to avoid digestive disorders, the feeding should be done at regular intervals.
15. Always avoid sudden change in the diet as it upsets whole stomach resulting in digestive disturbances and reduction in productive performance i.e. egg production, weight gain etc. the change if needed must always be gradual.
16. For better digestibility the animals should get at libitum the clean and fresh supply of drinking water.

Steps in ration formulation:

1. Identify animals to be fed by age, weight, function and specific condition (environmental) under which they are fed.
2. Select a set of nutrients requirement or allowance. Most appropriate for the animals being fed by using NRC or BIS specifications.

3. Select suitable ingredients to ensure that the ration is nutritionally balanced, palatable, safe and economical. You must aware about the:
   a. Chemical Composition and Nutritive value of common feed ingredients.
   b. Maximum level of inclusion in the diet to obtain optimum performance. Maximum level of inclusion of ingredient is dictated by:
      - Presence of anti nutritional factors (s).
      - Cost of ingredients.
      - Deficiency of nutrients, proper storage and shelf life of feed.
      - Its capacity to induce imbalance of nutrients and reduce the performance of animals.
4. Determine the necessary fixed amount of certain ingredients (minerals & vitamins) and then mix grains (s) relative to protein supplements to provide the desired level of amino acids.

Compounding of feeds

Compound feeds are feedstuffs that are blended from various raw material and additives. These feeds are formulated according to the specific requirements of the animals. They are manufactured by feed compounders as meal type, pellets or crumbles. The basic aim of compounding of feed is:

a. To use different feed ingredients efficiently in feeding of animals.
b. To produce uniform mixture of feeds with required nutrients.
c. To avoid selectivity and wastage of feeds.

Principles of compounding of feed:

Mixing of various feed ingredients in a single compound feed or in the form of a balanced ration in homogenous mixture is called compounding and mixing of feeds. The basic principle behind compounding of feeds is evenly distribution of the nutrients in the compounded feed i.e. any sample of the mix should represent the whole mixture in terms of levels of energy, protein, all vitamins and minerals for which the lot has been manufactured. This can be achieved by proper selection of ingredients, grinding and thorough mixing.

Premixing:

The recommended vitamin, trace minerals and feed additive are generally added in very small amount in ration. The proper mixing and distribution of these micronutrients in feed is achieved through premixing. Premix is formed by adding a suitable extender, which is a small part of the feed material to be mixed in the diet as per feed formula (preferably that is larger in quantity and non-reactive, like cereals). For this purpose, all the supplements and feed additives are added in about one kg of material, mixed thoroughly with the hands (with gloves) add more material and mix properly. Continue this process till it becomes about 10 kg then mix it with whole lot.

Mixing of liquids:

1. **Molasses:** Addition of molasses is difficult as it create problem of lumps formation. A good absorber like hominy feed, oat meat, de-oiled rice bran and wheat bran are selected for mixing molasses and then it is mixed as such in feed.
2. **Edible oil and fats:** These are used to enhance energy content of the mixed feed. Addition of fat/oil is difficult as it cause lumps formation. Fats create problem in pelleting especially when more than 5% fat is added in diet. This can be overcome by spraying of more than half of added fat on hot pellets. This prevents the crumbling of pellets.

Mixing of Feeds:

The purpose of mixing is to produce feed in which nutrients and medication are uniformly distributed. For an animal to grow, reproduce or lactate and remain in good health, it must receive a balanced ration that supplies nutrients and feed additives at proper concentrations. A satisfactory mixing process produces a uniform feed in minimum time with a minimum cost of overhead, power and labor. Incomplete mixing may occur because of inadequate or excessive mixing times, worn equipments and the use of ingredient which are difficult to mix (e.g. molasses, large fibrous or small dense particles). For proper mixing of different feed ingredients, mainly two types of mixers are used i.e. batch mixer and continuous flow mixer. The batch mixer is either vertical or horizontal. The vertical batch mixer is less expensive and mainly used at farm mills or small scale feed mills because of slow mixing (15-20 minutes). The horizontal batch mixer is used at large scale feed mills because mixing is fact (2-15 minutes) and it can accommodate mixtures of chopped hay, silage grains, molasses and other ingredients differing in particle size, density and texture. In continuous flow mixer all the ingredients are continuously mixed in the proper ratio to produce the desired diet. Speed of shaft is very high. It can properly mix more liquid ingredients as compared to horizontal batch mixer but bulky ingredients such as hay and silage cannot be mixed.

CHAPTER 30

COMPUTATION OF POULTRY RATION

Essential information required for the poultry rations formulation:

1. Nutrient requirements

The requirement for a given nutrient is the minimum quantity of that nutrient, when all other nutrients are supplied in adequate quantities, which will maintain normal growth and reproduction and at the same time prevent the development of nutritional deficiency symptoms. The accurate way to express the nutrient requirements is in amount /day /animal. Poultry are fed in groups and therefore, it is not practicable to express the requirements in amount/day/bird. The nutrient requirements can be expressed in relation to energy and protein content of the diet. NRC and ARC have published the minimum requirements for poultry. In India, BIS laid some specification on the nutrients requirements of poultry.

2. Nutrients composition of feed ingredient

The nutrients content in a feed ingredient depends on several factors. The most efficient way to provide nutrients for the birds is to analyze the feed ingredients for various nutrients and formulate the diets utilizing the obtained values.

3. Level of inclusion of feed ingredients

Each ingredient has its maximum level of inclusion in the diet. Inclusion of an ingredient beyond the maximum level may induce nutrients imbalance, difficulty in feed formulation and may reduce the performance of birds due to presence of antinutritional factors beyond tolerance level. Maximum level of inclusion of feed ingredients in poultry diets is presented in Table 1.

4. Availability and cost of feed ingredients

Feed represents the major cost of poultry production, constituting up to 70 % of the total. Feed cost, about 95 % is used to meet energy and protein requirements, about 3 to 4 % for major mineral, trace mineral and vitamin requirements, and 1 to 2 percent for various feed additives. Purchase and storage of feed ingredients during their season of production will helps in reducing cost of ingredient throughout the year. Least cost feed formulated with the help of computer. In computer feed formulation many nutrients can be taken into consideration at a time, however, by calculation, only a few nutrients can be considered. These include energy protein, amino acids, minerals and vitamins based on practical considerations.

Nutrient requirements for broilers and layers (BIS, 2024)

Table No. 1

S. No.	Characteristic	Requirement for														
		Broiler Feed			Layer Feed				Broiler Breeder Feed				Layer Breeder Feed			
		Starter	Grower	Finisher	Chick	Grower/ Pre- Layer	Layer Phase-I	Layer Phase - II	Chick	Grower/ Pre- Layer	Layer	Male	Chick	Grower/ Pre- Layer	Layer	Male
(1)	(2)	(3)	(4)	(5)	(6)	(7)	(8)	(9)	(10)	(11)	(12)	(13)	(14)	(15)	(16)	(17)
i)	Moisture, percent by mass, *Max*	11.0	11.0	11.0	11.0	11.0	11.0	11.0	11.0	11.0	11.0	11.0	11.0	11.0	11.0	11.0
ii)	Crude protein (N*6.25), Percent by mass, *Min*	22.5	21.0	19.5	20.0	16.0	18.0	16.0	20.0	16.0	16.0	14.0	20.0	16.0	17.0	16.0
iii)	Ether extract, percent by mass, *Min*	3.0	3.5	4.0	2.0	2.0	2.0	2.0	2.5	2.5	2.5	2.5	2.0	2.0	2.0	2.0
iv)	Crude fibre, Percent by mass, *Max*	5.0	5.0	5.0	7.0	9.0	9.0	10.0	7.0	9.0	9.0	9.0	7.0	9.0	9.0	9.0
v)	Acid insoluble ash, percent by mass, *Max*	2.5	2.5	2.5	4.0	4.0	4.0	4.5	4.0	4.0	4.0	4.0	2.5	2.5	2.5	2.5
vi)	Salt (as NaCl), Percent by mass, *Max*	0.5	0.5	0.5	0.5	0.5	0.5	0.5	0.5	0.5	0.5	0.5	0.5	0.5	0.5	0.5

Table No. 2: Requirements for Chicken Feeds (BIS, 2024)

S. No.	Characteristic	Requirement for														
		Broiler Feed			Layer Feed				Broiler Breeder Feed				Layer Breeder Feed			
		Starter	Grower	Finisher	Chick	Grower/ Pre-Layer	Layer Phase-I	Layer Phase-II	Chick	Grower/ Pre-Layer	Layer	Male	Chick	Grower/ Pre-Layer	Layer	Male
(1)	(2)	(3)	(4)	(5)	(6)	(7)	(8)	(9)	(10)	(11)	(12)	(13)	(14)	(15)	(16)	(17)
i)	Calcium (as Ca), percent by mass, *Min*	1.0	1.0	1.0	1.0	1.0 (Grower) 2.0 (Pre-layer)	3.75	4.0	1.0	1.0 (Grower) 2.0 (Pre -layer)	3.5	1.0	1.0	1.0 (Grower) 2.0 (Pre-Layer)	3.8	1.0
ii)	Total phosphorus, % by mass, *Min*	0.8	0.75	0.70	0.7	0.65	0.65	0.65	0.7	0.7	0.7	0.7	0.65	0.6	0.6	0.6
iii)	Available phosphorus, percent by mass, *Min*	0.48	0.46	0.44	0.45	0.40	0.35	0.33	0.45	0.45	0.40	0.40	0.45	0.40	0.35	0.40
iv)	Lysine, % by mass, *Min*	1.3	1.2	1.0	1.0	0.7	0.7	0.65	1.0	0.8	0.85	0.8	0.95	0.7	0.7	0.8
v)	Methionine, % by mass, *Min*	0.5	0.5	0.45	0.4	0.35	0.35	0.3	0.45	0.4	0.45	0.4	0.4	0.4	0.4	0.4
vi)	Methionine +Cysteine, % by mass, *Min*	0.96	0.92	0.85	0.65	0.59	0.65	0.61	0.52	0.50	0.55	0.50	0.65	0.59	0.65	0.65
vii)	Threonine, % by mass, *Min*	0.86	0.78	0.74	0.68	0.58	0.52	0.52	0.70	0.60	0.52	0.48	0.68	0.58	0.52	0.52
viii)	Metabolizable energy (Kcal/ kg), *Min*	3000	3050	3100	2800	2500	2600	2400	2800	2650	2700	2750	2800	2600	2600	2600

Table No. 3: Requirements for Minerals, Vitamins, Amino Acids and Fatty Acids per kg of Chicken Feeds

Sl No.	Characteristic	Requirement for														
		Broiler Feed			Layer Feed				Broiler Breeder Feed				Layer Breeder Feed			
		Starter	Grower	Finisher	Chick	Grower	Layer Phase-I	Layer Phase- II	Chick	Grower	Layer	Male	Chick	Grower	Layer	Male
(1)	(2)	(3)	(4)	(5)	(6)	(7)	(8)	(9)	(10)	(11)	(12)	(13)	(14)	(15)	(16)	(17)
i)	Man-ganese, mg/kg, *Min* mg/kg,*Min*	100.0	100.0	100.0	70.0	60.0	60.0	60.0	100.0	100.0	100.0	100.0	100.0	100.0	100.0	100.0
ii)	Iodine, mg/kg, *Min*	1.2	1.2	1.2	1.0	1.0	1.0	1.0	1.5	1.5	1.5	1.5	1.5	1.5	1.5	1.5
iii)	Iron, mg/kg, *Min*	80.0	80.0	80.0	70.0	60.0	60.0	60.0	80.0	80.0	80.0	80.0	80.0	80.0	80.0	80.0
iv)	Zinc, mg/kg, *Min*	80.0	80.0	80.0	60.0	60.0	60.0	60.0	80.0	80.0	80.0	80.0	80.0	80.0	80.0	80.0
v)	Copper, mg/kg, *Min*	12.0	12.0	12.0	12.0	9.0	9.0	9.0	12.0	12.0	12.0	12.0	12.0	12.0	12.0	12.0
vi)	Selenium, mg/kg, *Min*	0.15	0.15	0.15	0.15	0.15	0.15	0.15	0.15	0.15	0.20	0.15	0.15	0.15	0.20	0.15
vii)	Sodium, mg/kg, *Min*	0.19	0.19	0.14	0.18	0.18	0.15	0.15	0.20	0.17	0.15	0.15	0.18	0.18	0.15	0.15

Sl No.	Characteristic	Requirement for														
		Broiler Feed			Layer Feed				Broiler Breeder Feed				Layer Breeder Feed			
		Starter	Grower	Finisher	Chick	Grower	Layer Phase-I	Layer Phase- II	Chick	Grower	Layer	Male	Chick	Grower	Layer	Male
(1)	(2)	(3)	(4)	(5)	(6)	(7)	(8)	(9)	(10)	(11)	(12)	(13)	(14)	(15)	(16)	(17)
viii)	Chloride, mg/kg, *Min*	0.19	0.19	0.14	0.18	0.18	0.15	0.15	0.18	0.15	0.13	0.13	0.18	0.18	0.15	0.15
ix)	Potassium, mg/kg, *Min*	0.59	0.59	0.59	0.53	0.50	0.5	0.50	0.53	0.50	0.42	0.42	0.53	0.50	0.5	0.5
x)	Vitamin A, IU/kg, *Min*	11 000	11 000	11 000	9000	8000	8000	8000	12 000	12 000	15 000	12 000	12 000	12 000	15 000	12 000
xi)	Vitamin D3, IU/kg, *Min*	3000	3000	3000	1800	1600	1600	1600	2500	2500	3000	2500	2500	2500	3000	2500
xii)	Vitamin B1,mg/kg, *Min*	2.5	2.5	2.5	2.0	1.5	1.5	1.5	2.0	2.0	3.0	2.0	2.0	2.0	3.0	2.0
xiii)	Vitamin B2,mg/kg, *Min*	6.0	6.0	6.0	6.0	5.0	7.0	7.0	5.0	5.0	6.0	5.0	5.0	5.0	6.0	5.0
xiv)	Pantothenic acid, mg/kg, *Min*	15.0	15.0	15.0	10.0	9.0	9.0	9.0	15.0	15.0	25.0	15.0	15.0	15.0	25.0	15.0
xv)	Niacin, mg/kg, *Min*	40.0	40.0	40.0	40.0	20.0	20.0	20.0	40.0	40.0	50.0	40.0	40.0	40.0	50.0	40.0

Sl No.	Characteristic	Requirement for														
		Broiler Feed			Layer Feed				Broiler Breeder Feed				Layer Breeder Feed			
		Starter	Grower	Finisher	Chick	Grower	Layer Phase-I	Layer Phase- II	Chick	Grower	Layer	Male	Chick	Grower	Layer	Male
(1)	(2)	(3)	(4)	(5)	(6)	(7)	(8)	(9)	(10)	(11)	(12)	(13)	(14)	(15)	(16)	(17)
xvi)	Biotin, mg/kg, *Min*	0.15	0.15	0.15	0.10	0.10	0.10	0.10	0.20	0.20	0.20	0.20	0.20	0.20	0.20	0.20
xvii)	Vitamin B12,mg/ kg, *Min*	0.015	0.015	0.015	0.010	0.008	0.008	0.008	0.025	0.025	0.030	0.025	0.025	0.025	0.030	0.025
xviii)	Folic acid, mg/kg, *Min*	1.0	1.0	1.0	1.0	0.50	0.50	0.50	3.0	3.0	4.0	3.0	3.0	3.0	4.0	3.0
xix)	Choline, mg/kg, *Min*	500.0	500.0	500.0	500.0	200.0	400.0	400.0	850.0	850.0	700.0	500.0	850.0	850.0	700.0	500.0
xx)	Vitamin E,mg/kg, *Min*	30.0	30.0	30.0	15.0	10.0	10.0	10.0	20.0	20.0	50.0	20.0	20.0	20.0	50.0	20.0
xxi)	Vitamin K3, mg/ kg, *min*	1.5	1.5	1.5	1.5	1.5	1.5	1.5	2.0	2.0	3.0	2.0	2.0	2.0	3.0	2.0
xxii)	Vitamin B6, mg/ kg, *min*	5.0	5.0	5.0	3.0	3.0	3.0	3.0	5.0	5.0	6.0	5.0	5.0	5.0	6.0	5.0
xxiii)	Linoleic acid, %by mass, *Min*	1.1	1.1	1.1	1.0	1.0	1.0	1.0	1.0	1.0	1.0	1.0	1.0	1.0	1.0	1.0

Table 4: Maximum level of inclusion of feed ingredients in poultry diets

Ingredients	Inclusion level (%)	Ingredients	Inclusion level (%)
Energy source		Mustard meal	10
Maize	60	Maize–gluten feed	15
Wheat	50	Maize-gluten meal	15
Barley	20-40	Maize germ/meal	15
Oats	10-20	Niger meal	10
Bajra	50	Fish meal	10
Rice	40	Fish scrap	5
Rice bran	10-20	Fish solubles	3
Rice polish	25-40	Meat meal	10
Rice polish (deoiled)	10-20	Meat cum bone meal	5
Wheat bran	10-15	Blood meal	3
Salseed meal (deoiled)	3-5	Poultry byproduct meal	5
Molasses	5-10	Poultry hatchery byproduct meal	3
Animal and vegetable fat	10	Silkworm pupae meal	6
Poultry –manure meal	10	Feather meal	2
Tapioca flour	10-20	Skim milk dried	3
Protein Source		Penicillin mycelium waste	15
Groundnut-meal	40	**Leaf meals**	
Groundnut- meal (deoiled)	20	Alfalfa leaf meal	5
Soybean meal	40	Groundnut leaf meal	5
Sesame meal	20	**Fermentation byproducts**	
Safflower meal	20	Distillers dried solubles	10
Sunflower meal	20	Distillers dried grains	10
Linseed meal	4	Yeast, dried	5
Linseed meal	20	Yeast sludge, dried	10
Coconut meal	5	Mahua-cake	8
Cottonseed meal, de-gossypolized	5		

Steps in feed formulation:

The ration formulation should be carried out in a step-wise manner. For preparation of 100 kg ration for broiler starter (as per BIS 1992; CP= 23% and ME= 2800 Kcal/kg diet) the steps are as follows:

Step I: Minor ingredients are fixed or slack space may be left to include them later. Trace mineral vitamins, feed additives can be fixed because the contribution of major feed ingredients for these nutrients is little. Slack space may be left for addition of salt. Calcium and phosphorus sources, supplemental amino acids and fat at a later stage to balance the ration.

Step II: The levels of animal protein sources can be fixed. These can be added to fixed level because of the cost. These ingredients provide the limiting amino acids like lysine, cystine and methionine. Fish meal = 6.00 kg; Meat meal = 6.00 kg.

Step III: The level of cereal byproducts like rice polish, rice bran, wheat bran if to be added may be fixed. Alternatively the cereal byproducts and cereals can be fixed in certain ratio and can be added. Rice polish = 10.00 kg.

Step IV: Vegetable protein sources and energy sources are added to provide the required amount of protein. Till now 27.00 kg of ingredients were added (including the slack space) and these contributed 7.23 percent protein. Still 73.00 kg of ingredients are to be added and 15.77% protein is to be provided because the required CP in the diet should be 23.00%.

Groundnut meal and sesame meal in equal proportion as vegetable protein sources and maize as energy sources are considered. The required protein level can be calculated by Pearson's formula square method.

Pearson's square method:

Total quantity of ingredients	= 73 kg
Protein	= 15.77 kg
Protein as percent	$= \frac{15.77 \times 100}{73} = 21.60\%$

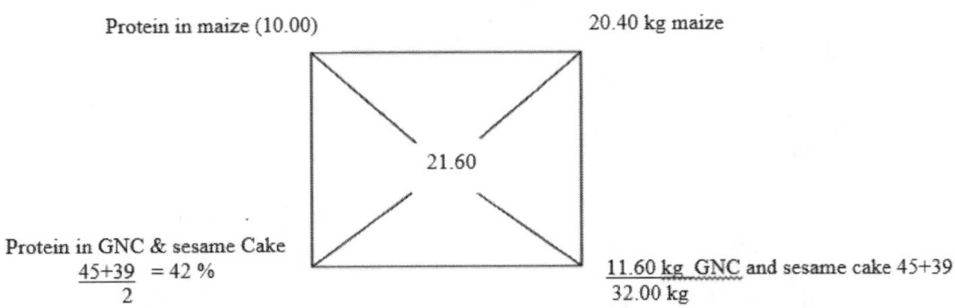

The desired protein level as percent is entered in the centre of the square. The protein percent of vegetable protein source (groundnut cake and sesame cake) and maize is entered at the corners. The difference between the protein percent is the ratio of different ingredients to be mixed.

For 73 kg maize	= 20.40/32.00 × 73	= 46.54 kg
For vegetable protein (1:1 GNC+ Sesame)	= 11.60 × 73	= 26.46 kg
Groundnut cake	= 26.46/2	= 13.23 kg
Sesame cake	= 26.46/2	= 13.23 kg

Step V: The ME content of the diet has to be balanced. The shortfall of 124.69 Kcal (2800-2675.31) of ME can be met by supplementation of animal fats. However the addition of fat is the question of economics and may not be practical in all circumstances.

Step VI: The phosphorus content of the diet is calculated in terms of available phosphorus. Phosphorus from animal and inorganic sources is considered completely available whereas that from plant sources is considered to be 30% available.

Step VII: The requirements of calcium can be met by supplementation of limestone and bone meal.

Step VIII: The limiting amino acids in synthetic from can be supplemented to meet the requirement for these limiting amino acids can be met by increased level of inclusion of animal protein sources.

Step IX: A check is made for the total of the ingredients and also for all the nutrients if desired. If the total is still below 100, cereals or cereal byproducts can be added to make it 100.

Table 5: Examples of Broiler Starter Diet-1

Ingredients	Parts (Kg)	CP (%)	ME (Kcal/kg)	Ca	P	Lysine	Met	Cys	CP (%)	ME (Kcal/kg)	Ca	P	Lysine	Met	Cys
Fish meal	6.00	43.1	1834	7.16	1.67	4.17	1.42	0.57	2.59	110.04	0.43	0.10	0.25	0.09	0.03
Meat meal	6.00	56.2	2319	2.68	2.06	4.00	0.84	0.63	3.37	139.14	0.16	0.12	0.24	0.05	0.04
Rice polish	10.00	12.7	2937	0.27	1.37	0.44	0.24	0.09	1.27	293.70	0.14	0.14	0.04	0.02	0.01
Maize	46.54	10.00	3309	0.25	0.40	0.18	0.15	0.08	4.65	1540.00	0.17	0.19	0.08	0.07	0.04
Groundnut cake	13.23	45.00	2596	0.23	0.59	1.20	0.57	0.57	5.95	343.45	0.03	0.08	0.16	0.07	0.07
Sesame meal	13.23	39.10	1882	0.48	0.98	1.14	1.23	0.53	5.17	248.98	0.06	0.13	0.15	0.16	0.07
Animal fat	1.62	-	7700	-	-	-	-	-	-	124.69	-	-	-	-	-
Bone meal	0.50	-	-	29.0	13.6	-	-	-	-	-	0.15	0.07	-	-	-
Limestone	1.00	-	-	34.0	-	-	-	-	-	-	0.34	-	-	-	-
Salt	0.35	-	-	-	-	-	-	-	-	-	-	-	-	-	-
Lysine HCl													0.15		
Trace minerals															
Vitamins															
Coccidiostat	1.53	-	-	-	-	-	-	-	-	-	-	-		-	-
Arsenicals															
Antibacterial agent															
Total									23.00	2800.00	1.48	0.83	0.92		

Table 5: Examples of Broiler Starter Diet-2

Ingredient	Quantity (%)	Category	Nutrient Contribution
Maize	35.0	Cereal Grain	Energy
Sorghum	5.0	Cereal Grain	Energy
Wheat	7.0	Cereal Grain	Energy,
Broken Rice	3.0	Cereal Grain	Energy
Barley	2.0	Cereal Grain	Energy, Fiber
Soybean Meal (46% CP)	25.0	Protein Source	Protein
Groundnut Cake	5.0	Protein Source	Protein
Cottonseed Meal (41% CP)	4.0	Protein Source	Protein
Sunflower Meal (34% CP)	3.0	Protein Source	Protein
Fish Meal (60% CP)	2.5	Animal Protein Source	Protein, Minerals
Meat and Bone Meal (50% CP)	2.0	Animal Protein Source	Protein, Calcium, Phosphorus
Soy Oil	3.5	Fat Source	Energy
Rice Bran Oil	1.5	Fat Source	Energy
Dicalcium Phosphate	1.6	Mineral Source	Calcium, Phosphorus
Limestone	1.2	Mineral Source	Calcium
Salt	0.3	Mineral Source	Sodium, Chlorine
Vitamin Premix	0.1	Vitamin Source	Vitamins
Mineral Premix	0.1	Mineral Source	Trace Minerals
DL-Methionine	0.2	Amino Acid Supplement	Methionine (Amino Acid)
L-Lysine	0.2	Amino Acid Supplement	Lysine (Amino Acid)
Threonine	0.1	Amino Acid Supplement	Threonine (Amino Acid)
Toxin Binder	0.1	Feed Additive	Mycotoxin Management
Coccidiostat	0.2	Feed Additive	Coccidiosis Prevention
Enzyme (Phytase)	0.1	Feed Additive	Phosphorus Utilization
Choline Chloride	0.1	Vitamin Supplement	Choline
Total	100.0		

Nutrient Composition

Nutrient	Value (%)
Metabolizable Energy	3050 kcal/kg
Crude Protein	22.5
Crude Fiber	3.8
Ether Extract (Fat)	5.5

Nutrient	Value (%)
Calcium	0.95
Available Phosphorus	0.45
Sodium	0.15
Methionine	0.50
Methionine + Cystine	0.90
Lysine	1.25
Threonine	0.82
Tryptophan	0.24
Isoleucine	0.92
Valine	1.02
Arginine	1.38
Leucine	1.75
Histidine	0.36
Phenylalanine	1.05
Glycine + Serine	1.85

Table 6: Ration for Replacement Pullets and Layers

Feed Ingredients (kg, 100 kg ration)	Chick (0-8 weeks)		Grower (8-20 weeks)		Layer/ Breeder (over 20 weeks)	
	I	II	I	II	I	II
Maize or maize and bajra / wheat / white jower (50:50)	32.0	10.0	22.0	10.0	32.0	13.0
GNC/ soybean meal or GNC and till cake (50:50)	21.5	9.0	20.5	7.5	13.5	7.5
Guar meal and mustard cake (50:50) or sunflower cake	8.0	8.0	8.0	8.0	8.0	8.0
Fishmeal or fishmeal and meal (50:50)	8.0	10.0	6.0	6.0	5.0	5.0
Molasses	-	-	5.0	-	5.0	-
Rice polish	28.0	40.5	36.0	40.0	30.0	20.0
Deoiled rice polish	-	10.0	-	10.0	-	20.0
Wheat bran	-	-	-	10.0	-	10.0
Maize gluten meal	-	10.0	-	6.0	-	10.0
Bone meal (Steamed)	-	-	-	-	1.0	1.0
Lime stone					3.0	3.0
Mineral mixture	2.5	2.5	2.5	2.5	2.5	2.5
Feed supplements*	0.13	0.13	0.13	0.13	0.13	0.13

*** Feed supplements per 100 kg of mesh as 0.05 kg of bifuran, 0.06 kg of Aurofac -10 (cynamide) and 0.02 kg of Vitablend (Glaxo). Bifuran and Aurofac are not required in layer/breeder rations.**

CHAPTER 31

CALCULATION OF NUTRIENT REQUIREMENTS FOR BROILER BIRDS

Nutrient requirement are the guidelines to achieve maximum returns with optimum inputs. These requirements vary within species and depend on many factors which ultimately affect the maintenance and growth of meat producing birds.

Objectives:

1. To calculate the protein requirement of meat type birds.
2. To calculate the energy requirement of meat type birds.

Calculation of protein requirement

Requirement of protein for growth of meat type birds can be categorized into three parts:

1. Protein required for maintenance.
2. Protein required for feather growth.
3. Protein required for tissue growth.

Note:

1. Energy requirements in poultry is expressed in terms of metabolizable (ME)
2. Efficiency of protein utilization for maintenance, feather and tissue growth in meat type birds is 64 percent.)
3. Meat type birds require energy for maintenance, growth, regulation of body temperature and activity.
4. For newly hatched chicken weighing approximately 40 g needs 8 Kcal ME/ day for maintenance.

5. The energy for growth range between 1.5 to 3.0 Kcal per gram of body weight gain.
6. The endogenous N loss in chicken is found to be 250 mg N/Kg BW/day or in other words it is 0.0016 g crude protein /gain in weight/day.
7. At 3 weeks of age feather represents 4 percent of body weight and after 4th week feather represents 7 percent of body weight.
8. On an average chicken carcass contains 18 percent crude protein and feather contains 82 percent crude protein.

Daily protein requirement for growing meat type birds can be calculated as:

$$\text{For Maintenance} = \frac{\text{Body weight} \times 0.0016}{0.64}$$

$$\text{For Feather growth} = \frac{0.04 \times \text{Daily weight gain} \times 82}{64}$$

$$\text{For tissue growth} = \frac{\text{Daily weight gain} \times 18}{64}$$

CHAPTER 32

FORMULATION OF RATIONS FOR BROILER BIRDS

Meat type birds (Broilers) grow rapidly and can be marketed when about 5 to 6 weeks of age. They usually attain 1.5 to 2.0 kg live weight by this time. Boiler rations are formulated in such a way that they promote early, rapid growth. High protein and high energy rations are formulated for meat type birds.

Objectives:

To formulate rations for Starter and Finisher phases of Broiler birds.

Procedure:

The nutrient requirements for meat type birds are presented in the following table as per BIS (2007). BIS has divided meat type birds into two phases: Broiler pre-starter, starter and finisher phases. Suggested levels as per the requirements of the improved WLH strains are also included in the Table 1.

Table 1: Nutrient requirement for different categories of broilers (BIS-2007)

Characteristic	Broiler Pre- starter	Broiler Starter	Broiler Finisher
Moisture,% (Max.)	11.00	11.00	11.00
Crude Protein, % (Min.)	23.00	22.00	20.00
Crudefibre, % (Max.)	5.00	5.00	5.00
Acidinsolubleash, % (Max.)	2.50	2.50	2.50
Salt,% (Max.)	0.50	0.50	0.50
Calcium, % (Min.)	1.0	1.0	1.0
Phosphorous (Available), % (Min.)	0.45	0.45	0.45
LinoleicAcid, % (Min.)	1.00	1.00	1.00
Lysine,% (Min.)	1.30	1.20	1.00

Methionine, % (Min.)	0.50	0.50	0.45
Meth.+cystine,%	0.90	0.90	0.85
Metabolizable Energy (Kcal/ Kg) Min.	3000	3100	3200
Ether extract, % (Min.)	3.0	3.5	4.0
Minerals and Vitamins:			
Manganese, mg/kg	100.00	100.00	100.00
Iodine, mg/kg	1.20	1.20	1.20
Iron, mg/kg	80.00	80.00	80.00
Zinc, mg/kg	80.00	80.00	80.00
Copper, mg/kg	12.00	12.00	12.00
Selenium, mg/kg	0.15	0.15	0.15
VitaminA, IU/Kg	6000	6000	6000
VitaminD_3, IU/Kg	600	600	600
Thiamin, mg/kg	5.00	5.00	5.00
Riboflavin, mg/kg	6.00	6.00	6.00
Pantothenic acid, mg/kg	15.00	15.00	15.00
Nicotinic acid, mg/kg	70.00	70.00	70.00
Biotin, mg/kg	0.20	0.20	0.20
VitaminB_{12}, mg/kg	0.015	0.015	0.015
FolicAcid, mg/kg	1.00	1.00	1.00
Choline, mg/kg	1400	1400	1000
VitaminE, mg/kg	15.00	15.00	15.00
VitaminK, mg/kg	1.00	1.00	1.00
Pyridoxine, mg/kg	5.00	5.00	5.00

Different feed ingredients have some limitations, so they can be included in rations of different categories of meat type birds at certain levels. Maximum inclusions of various feed ingredients are shown in table.

Table 2: Inclusion levels of various feed ingredients in the rations

S. No.	Ingredients	Chicks & Broilers
	Cereals	
1	Maize	70
2	Bajra	30
3	Wheat	20
4	Rice kani	10
5	Barly	10

	Cereal by product	
6	Rice polish	20
7	De-oiled Rice Polish	5
8	Wheat Bran	5
	Vegetable proteins	
9	Soybean meal	35
10	Full fat soybean meal	10
11	Groundnut Ext.	35
12	Sunflower cake	10
13	Mustard cake	3
14	De-oiled Mustard cake	
15	Cotton seed meal	5
16	Guar meal	3
17	Maize gluten meal	10
18	Brewer's dried grains	
	Animals proteins	
19	Fish meal	10
20	Meat meal	5
21	Meat cum bone meal	5
22	Leather meal	3

It is very important to maintain calorie: protein (Kcal/% protein) ratio in the diet. Calorie: protein ratio influences the feed consumption and feed efficiency. If the energy content of feed is increased, feed intake reduces which ultimately reduces the intake of other nutrients. If the energy content of feed is increased it is suggested to increase the nutrients proportionally.

Pearson square method and Hit and Trial methods are used for formulating rations.

CHAPTER 33

CALCULATION OF NUTRIENT REQUIREMENTS FOR EGG TYPE BIRDS

The poultry birds start laying eggs at about 18-20th week of age. The laying phase of egg type poultry birds is critical and can be divided into two phases. Stage -1 is from 20th week to 40th week of age. This is the phase when egg production starts and reaches at peak production. The body and egg size also increases in this phase.

Objectives:

1. To calculate the protein requirement of egg laying birds.
2. To calculate the energy requirement of egg laying birds.

Calculation of protein requirement

During production cycle daily of protein is dependent upon the following states:

a. Maintenance

b. Growth

c. Egg production

Note:

1. Efficiency of protein utilization for maintenance, growth and egg production in egg type birds is 55 %.
2. Daily total endogenous N excretion of adult egg type layer can be calculated using the equation -200 $W^{0.75}$ and expressed as mg N/day.
3. On an average egg contains 12 % protein.

Daily protein requirement for egg laying birds can be calculated as:

$$\text{For maintenance} = \frac{200\,W^{0.75} \times 6.25}{0.55}$$

For growth = (Average weight at the time of calculation – average weight at the start of egg production) = X

$$\text{For Growth} = \frac{X}{\text{No. of days}} = Y$$

$$\text{So, } \frac{Y \times 0.18}{0.55}$$

$$\text{For egg production} = \frac{\text{Average weight of Egg} \times 0.12}{0.55}$$

Total of above three will give the daily protein required for laying hen.

Calculation of energy requirement:

Since chicken has higher body temperature than mammals so avian species has greater maintenance energy requirement. Experimental studies have shown that ME requirements are approximately 18 percent higher than net energy requirements. The basal metabolism studies indicate that the NEm of adult hen can be expressed as: 83 $W^{0.75}$ Kcal per day = X

Therefore,

$$\text{NEm} = \frac{X}{0.82} = Y$$

Activity allowance of energy for hen reared in deep litter system is taken as 50 % of MEm and if the birds are reared in cages where movement is restricted the activity allowance for caged birds is taken as 37 %.

So total MEm requirement for maintenance is Y+ (37 % of Y)

Energy content of large egg is 86 Kcal

If the percent egg production is 90%

The MEm for egg = 86 x0.90

CHAPTER 34

FORMULATION OF RATIONS FOR EGG TYPE BIRDS-STARTER AND GROWER

The purpose of rearing pullets differs from the production of broilers. In broilers every effort is made obtain faster growth with desirable carcass composition. However, diets for replacement pullets should contain all nutrients needed for normal growth and development. Diet should produce good feathering and provide ample storage of nutrients.

Objectives:

1. To formulate rations for egg type starter birds.
2. To formulate rations for egg type grower birds.

Procedure:

1. The main purpose of rearing replacement pullets is to produce hens with a potential for maximum production.
2. During starter phase (0-8 weeks) need for growth of tissues and feather are most important. However, during grower (9-18 weeks) phase the maintenance needs contribute much more towards nutrient requirements.
3. Choose the feed ingredients like de-oiled rice bran which provide low energy but higher fiber contents.
4. If possible use combinations of different cereal and cereal by products
5. Combination of different vegetable protein sources along with animal protein sources can be incorporated in the diets of starter chicks.
6. The required level of calorie: Protein ratio during starter phase in egg type birds is 145 and during grower phase is 160.

7. Development of skeletal system and muscular system mainly takes place in grower phase. Feed consumption should be closely monitored so that required nutrient levels are met.
8. Shifting of feed from one phase to other phase should be gradual.
9. Calcium level should be increased from 1.0 to 2.5 percent after 15th week of age to develop medullary bone which acts as calcium reservoir.

The nutrients requirements for egg type birds are presented in the following table as per BIS (2007) and suggested levels during these phases as required with latest needs.

Table 1: Nutrients requirements for different categories of egg type birds (BIS 2007).

Characteristic	Chick Starter	Grower	Layer Phase-I	Layer Phase- II
Moisture, % (Max.)	11.00	11.00	11.00	11.00
Crude Protein, % (Min.)	20.00	16.00	18.00	16.00
Crude fibre, % (Max.)	7.00	9.00	9.00	10.00
Acid insoluble ash, % (Max.)	4.00	4.00	4.00	4.50
Salt, % (Max.)	0.50	0.50	0.50	0.50
Calcium, % (Min.)	1.00	1.00	3.00	3.00
Phosphorous (available), % (Min.)	0.50	0.50	0.50	0.50
Linoleic Acid, % (Min.)	1.00	1.00	1.00	1.00
Lysine, % (Min.)	1.00	0.70	0.70	0.65
Methionine, % (Min.)	0.45	0.35	0.35	0.30
Meth.+cystine,%	0.70	0.60	0.60	0.55
Metabolizable Energy(Kcal ME/Kg), Min.	2800	2500	2600	2400
Ether extract, % (Min.)	2.0	2.0	2.0	2.0
Minerals and Vitamins				
Manganese, mg/kg	90.00	50.00	55.00	55.00
Iodine, mg/kg	1.00	1.00	1.00	1.00
Iron, mg/kg	120.00	90.00	75.00	75.00
Zinc, mg/kg	60.00	50.00	75.00	75.00
Copper, mg/kg	12.00	9.00	9.00	9.00
VitaminA, IU/Kg	6000	6000	6000	6000
VitaminD3, IU/Kg	600	600	1200	1200
Thiamin, mg/kg	5.00	3.00	3.00	3.00
Riboflavin, mg/kg	6.00	5.00	5.00	5.00
Pantothenicacid, mg/kg	15.00	15.00	15.00	15.00
NicotinicAcid, mg/kg	70.00	60.00	60.00	60.00
Biotin, mg/kg	0.20	0.15	0.15	0.15

Characteristic	Chick Starter	Grower	Layer Phase-I	Layer Phase- II
Vitamin B_{12}, mg/kg	0.015	0.10	0.10	0.10
Folic Acid, mg/kg	1.00	0.50	0.50	0.50
Choline, mg/kg	1300	900	800	800
Vitamin E, mg/kg	15.00	10.00	10.00	10.00
Vitamin K, mg/kg	1.00	1.00	1.00	1.00
Pyridoxine, mg/kg	5.00	5.00	5.00	5.00

CHAPTER 35

FORMULATION OF RATIONS FOR LAYER BIRDS

Feed cost represents almost 70 percent of total cost of egg production. Formulating rations for layers with high efficiency of feed utilization becomes the main concern to encourage optimum growth and egg production. Body weight at the starts of laying is one of the most significant factor influencing early egg weight and total egg production. The Judicious use of feed ingredients to supply adequate nutrients to meet hens need becomes very important while formulating layer rations.

Objectives:

To formulate rations for layers.

Procedure:

1. Start layer ration as soon as first egg is laid.
2. To meet the increased requirement of calcium we must add 8-9 % limestone in the diets of Layers. 50% of the limestone should be of 3 to 4 mm in size and rest of the limestone should be used in powder form.
3. Do not give excess calcium them recommended levels as this can lead to the deficiency of phosphorus and will also affects the appetite.
4. Nutrients like lysine, methionine, linoleic acid and energy levels in the feed influences egg size. Bird eats to meet its energy requirement. Once energy requirement is full filed bird will not eat. High or low energy diet lead to reduced or increased feed intake which will alter the intake of other nutrients. Accordingly nutrients may be adjusted as per the available energy in the diet. Calorie: Protein ratio for laying stock should be 146.

Pearson square method and hit and trial methods are used for formulating rations. Pearson square method gives the idea that how much protein and energy source is required to be added in the rations. Level of inclusion, bioavailability of the feed ingredients, cost etc are many other factors which should be kept in mind while formulating rations. Many feed ingredients have some limitations so they can be added to limited extent in layer rations, like excess of fish meal create salt toxicity. Maximum inclusions levels of commonly used feeds are presented in the following table 1.

The Inclusion level of various feed ingredients in the rations of laying bird is suggested in the following Table 1.

Table 1: Inclusion levels of various feed ingredients in the layer rations.

S. No.	Ingredients	Grower & Layers
	Cereals	
1	Maize	70
2	Bajra	60
3	Wheat	30
4	Rice kani	20
5	Barley	20
	Cereal by product	
6	Rice polish	30
7	De-oiled rice polish (DORP)	30
8	Wheat bran	10
	Vegetable proteins	
9	Soybean meal (SBM)	25
10	Full fat soybean meal	10
11	Ground nut cake (GNC)	25
12	Sunflower cake	20
13	Mustard oil cake (MOC)	5
14	De-oiled mustard cake	
15	Cotton seed meal	10
16	Guar meal	5
17	Maize gluten meal	20
18	Brewer's dried grains	
	Animals proteins	
19	Fish meal	6
20	Meat meal	5
21	Meat cum bone meal	5
22	Leather meal	3

CHAPTER 36

CALCULATION OF NUTRIENT REQUIREMENTS FOR SWINE

Nutrient requirements are the statement of the amount of nutrients required by animals that should support normal body functions. Pigs require nutrients for maintenance, growth, production, reproduction and lactation.

The aim of study is -

1. Estimate the energy requirements for maintenance, growth, reproduction, lactation and work
2. Estimate the protein requirements for maintenance, growth, reproduction, lactation and work

Calculation of energy requirements

A. For maintenance: The maintenance energy requirement is the amount of food energy needed to keep the animal at a constant body weight and composition. It can be estimated by three methods; fasting metabolism, feeding trial method and comparative slaughter experiment. The ME requirement for maintenance (ME_m) includes the needs of all body functions and moderate activity. These requirements are usually expressed on a metabolic body weight basis, which is defined as body weight raised to the 0.75 power ($BW^{0.75}$). For pigs, the estimated maintenance energy requirement/kg of $BW^{0.75}$ is:

$$MEm = 106 \times BW^{0.75} \text{ (kcal/day)}$$

$$DEm = 110 \times BW^{0.75} \text{ (kcal/day)}$$

The body weight is in kilograms.

Calculate maintenance energy requirement of a pig weighed 100kg.

Ans $MEm = 106 \times BW^{0.75}$ kcal/day

$= 106 \times 100^{0.75}$ kcal/day

$$= 106 \times 31.6 \text{ kcal/day}$$
$$= 3351.7 \text{ kcal/day}$$

B. For Growth

The DE requirements of growing pigs kept in thermo neutral zone can be estimated by factorial method by using following equation

$$DE = DE_m + DE_{pr} + DE_{fr}$$

Where DE_m, DE_{pr} and DE_{tr} are requirements for maintenance , protein retention and fat retention, respectively. The daily maintenance requirement of the growing pigs is considered to be 110 kcal of DE/kg $BW^{0.75}$ (or 106 kcal of ME/kg $BW^{0.75}$). Estimates for the energy costs of protein retention (ME_{pr}) is 10.6 kcal of ME/g and that of fat deposition (ME_{fr}) is 12.5 kcal of ME/g (NRC, 1998). One kilogram of lean muscle tissue contains about 20 to 23 percent protein, whereas 1 kg of adipose tissue contains 80 to 95 percent fat. The requirement for ME is the requirement for DE ×0.96.

Calculate DE requirement of a pig weighed 100 kg kept in thermo neutral zone if 20 g of protein and 85 g of fat is daily retained in the body.

Ans $ME = ME_m + ME_{pr} + ME_{fr}$

$$ME = 106 \times 100^{0.75} + 10.6 \times 20 + 85 \times 12.5 \text{ kcal/day}$$
$$ME = 3351.7 + 212 + 1000 \text{ kcal/day}$$
$$ME = 4563.7 \text{ kcal/day}$$
$$DE = \frac{ME}{0.96} = \frac{4563.7}{0.96} \text{ kcal/day}$$
$$DE = 4753.8 \text{ kcal/day}$$

C. For pregnancy: The daily energy requirement is the sum of the requirements for maintenance, for protein and fat accretion, and for thermoregulation. Tissue accretion is the sum of that in the maternal body and the products of conception. The daily maintenance requirement of the gestating sow is considered to be 106 kcal of ME/kg $BW^{0.75}$. The energy cost of protein accretion is assumed to be 10.6 kcal of ME/g and that accretion to be 12.5 kcal of ME/g. The daily energy requirement for growth of the products of conception is 35.8 kcal of ME/fetus. Additional energy is required when sows are maintained in a cold environment. The model predicts that a sow with an average gestation weight of 200 kg will need approximately 240 additional kcal of ME (250 kcal of DE) per day for each 1°C below 20 °C. No adjustment is made for temperatures above 20 °C. The requirement for DE is the requirement for ME/0.96.

D. For lactation: Estimation of nutrient requirements for lactation sows is complicated by the sow's propensity to contribute energy and amino acids retrieved from her own body to help support her milk production. The daily energy requirement of lactating sow is the sum of that expended in maintenance plus the gross energy of milk less the contribution made by the mobilized body tissue. The daily maintenance requirement of the lactating sow is considered to be 106 kcal of ME/kg $BW^{0.75}$ (or 0.44 MJ of ME/kg $BW^{0.75}$), the same as for the gestating sow. The energy requirement for milk production can be estimated from the growth rate of the suckling pigs in the litter.

Milk energy= (4.92 x Litter gain)-(90 x Number of pigs)

Where milk energy is expressed in kcal GE/day and Litter gain is in g/day.

The amount of dietary ME required to produce this amount of milk energy is calculated by dividing the milk energy by 0.72, assuming that the marginal efficiency of use of ME for milk production is 72 percent. This estimate is at 20 ^{0}C, an additional 310 kcal of dietary ME (323 kcal of DE) will be consumed per day by sows for every 1^0rise or fall in temperature from 20^0C. The contribution towards milk production from sow's body can be calculated by the equation

Protein gain=1.47+ (0.0942 x ADG)

Where protein gain is in g/day and ADG as the sow's average daily gain of body weight in grams. Each grams of protein retrieved from the sow's body is assumed to provide 5.6 kcal of GE toward meeting the energy requirement. The amount of protein is divided by 0.23 to estimate the amount of lean tissue mobilized (assuming lean tissue is 23 percent protein). Subtracting the amount of body weight lost gives an estimate of the amount of adipose mobilized. This adipose tissue is considered to be 90 percent fat and it is assumed that 9.4 kcal of GE per gram of fat mobilized is available to be applied towards the energy requirement. The total energy from mobilized tissue is used with an efficiency of 0.88 to meet the energy demands of lactation.

Calculate ME requirement of a lactating sow with body weight 200 kg, litter size 8, litter gain 300 g/d, losing at rate of 100 g/d kept at thermo neutral zone.

Ans: Sum of energy expended in maintenance plus gross energy of milk less the contribution made by mobilized body tissue

$$ME_L = ME_m + ME_{milk} - ME_{mobilized\ tissue}$$

Energy for maintenance

$$ME_m = 106 \times 200^{0.75} = 5637.4 \text{ kcal/d}$$

Energy for milk

$$ME_{milk} = (4.92 \times \text{Litter gain}) - (90 \times \text{Number of pigs})/0.72$$

Because efficiency of use of ME for milk production is 72%

$$ME_{milk} = ((4.92 \times 300) - (90 \times 8))/0.72 = 1050 \text{ kcal/d}$$

Energy from mobilized body tissue

$$\text{Protein gain} = 1.47 + (0.0942 \times \text{ADG})$$

$$\text{Protein gain} = 1.47 + (0.0942(-100))$$

$$\text{Protein gain} = 1.47 - 9.42 = 8\ \text{g}$$

$$\text{Amount of energy supplied by protein} = 8 \times 5.6 = 44.8\ \text{g}$$

$$\text{Amount of lean tissue mobilized} = 8/0.23 = 34.8\ \text{g}$$

(Assuming lean tissue contains 23 percent protein)

$$\text{Amount of adipose tissue mobilized} = 100 - 34.8 = 65.2\ \text{g}$$

(Assuming adipose tissue contains 90 percent fat)

$$\text{Amount of Fat mobilized} = 65.2 \times 90/100 = 58.7\ \text{g}$$

$$\text{Amount of energy supplied by fat} = 58.7 \times 9.4 = 551.4\ \text{kcal}$$

$$\text{Total energy from protein and fat} = 44.8 + 551.4 = 596.6\ \text{kcal}$$

$$ME_{\text{mobilized tissue}} = 596.6 \times 0.88 = 485.6\ \text{kcal}$$

(as efficiency of utilization of energy from mobilized tissue for milk production is 88%)

$$ME_L = ME_m + ME_{milk} - ME_{\text{mobilized tissue}}$$

$$ME_L = 5637.4 + 1050 - 485.6\ \text{kcal/day}$$

$$ME_L = 6201.8\ \text{kcal/day}$$

Estimation of protein requirements

A. For maintenance: In general, the estimation of protein requirements for maintenance and growth can be done by factorial method, nitrogen balance and feeding trial method. For pig and poultry, the protein requirements are usually stated for maintenance and production together. The protein requirements in swine are discussed in terms of amino acids (NRC, 1998). The daily true ileal digestible lysine requirement for maintenance is assumed to be 36 mg/kg of metabolic body weight ($BW^{0.75}$) for swine. The ideal ratios of amino acids to lysine for maintenance, protein accretion, milk synthesis, and body tissue for swine are given in Table 1.

B. For growth: For growing finishing pigs, the daily lysine requirement is the sum of the requirements for maintenance and for protein accretion. The daily true ileal digestible lysine requirement for maintenance is assumed to be 36 mg/kg of metabolic body weight ($BW^{0.75}$). The daily amount of lysine needed to support protein accretion is the product of two numbers: (1) the daily amount of protein accreted and (2) the amount of true digestible lysine needed for each gram of protein accreted. The amount of true digestible lysine needed for each gram of protein accreted is 0.122. True ileal digestible lysine required for protein gain for pigs from 20 to 120 kg BW (g/day) is 0.12 × whole body protein gains. The true ileal digestible amino acid requirements (g/day) can be calculated from the true ileal digestible lysine required for maintenance is multiplied by the ratio of each amino acid to lysine for maintenance (Table 1). The true ileal digestible lysine for gain is multiplied by the ratio of each amino acid to lysine for gain (Table 1). The sum of requirement for maintenance and gain for each amino acid is the daily total true ileal digestible lysine requirement.

Table 1: Ideal ratios of amino acids to lysine for maintenance, protein accretion, milk synthesis, and body tissue for swine (NRC, 1998).

Amino acids	Maintenance	Protein accretion	Milk synthesis	Body tissue
Lysine	100	100	100	100
Arginine	-200*	48	66	105
Histidine	32	32	40	45
Isoleucine	75	54	55	50
Leucine	70	102	115	109
Methionine	28	27	26	27
Methionine + cysteine	123	55	45	45
Phenylalanine	50	60	55	60
Phenylalanine + tyrosine	121	93	112	103
Threonine	151	60	58	58
Tryptophan	26	18	18	10
Valine	67	68	85	69

*** The negative value for arginine reflects arginine synthesis in excess of the needs for maintenance.**

C. For pregnancy: Factorial method used to calculate the daily protein requirements for gestating sows. The average daily nitrogen retention is the sum of maternal nitrogen retention and nitrogen in the products of conception.

1. Maternal nitrogen retention

   N retention (g/day)= [(Maternal lean tissue)/115]×0.23×0.16

2. Nitrogen in the products of conception

   N retention (g/day)= No. of pigs ×0.34 g/day

3. True ileal digestible lysine requirement for gain is calculated from N retention

   True ileal digestible lysine (g/day)= (Total N retention ×0.807)

4. True ileal digestible lysine requirement for maintenance

   True ileal digestible lysine (g/day)= 0.036 ×metabolic BW

5. Total true ileal digestible lysine requirement is the sum of the requirement for gain and maintenance.
6. The true ileal digestible lysine required for maintenance (step 4) is multiplied by the ratio of each amino acid to lysine for maintenance (Table 1). The true ileal digestible lysine required for gain (step 3) is multiplied by the ratio of each amino acid to lysine for protein accretion (Table 1). The total requirement is the sum of the requirements for maintenance and gain.

D. For lactation: The amino acid requirements for lactating sows can be calculated by factorial approach. The digestible lysine required is

1. Apparent digestible lysine for milk production (g)

   = (Daily pig weight gain × pigs in litter ×0.022)-6.39
2. Apparent digestible lysine for milk production (g/day) is converted to a percentage using DE intake and DE concentration.
3. Percentage apparent digestible lysine is converted to percentage true digestible lysine

   True digestible lysine % = 1.050013 × apparant digestible lysine %+ 0.022052
4. Percentage true digestible lysine is converted to g/day using DE intake and DE concentration.
5. Lysine for maintenance = 0.036 × metabolic BW
6. Lysine from tissue = Change in protein × 0.065
7. Total true ileal digestible lysine requirement is sum of steps 4, 5, and 6.

True digestible amino acid requirements (g/day) of lactating sows can be calculated from true ileal digestible lysine. The true ileal digestible lysine required for maintenance (step 5) is multiplied by the ratio of each amino acid to lysine for maintenance (Table 1). The true ileal digestible lysine required for milk production is multiplied by the ratio of each amino acid to lysine in milk protein (Table 1). The lysine from tissue is multiplied by the ratio of each amino acid to lysine in tissue (Table 1). The total requirement is the sum of requirements for maintenance, milk production and tissue change.

CHAPTER 37

RATION FORMULATION FOR SWINE

Pig is the most prolific (10-12 piglets per farrowing) meat producing animal having very good feed conversion efficiency (1:3) and dressing percentage (60-80%). Pigs convert inedible feeds, forages, certain grain byproducts, meat byproducts, damaged feeds and garbage into valuable nutritious meat. Most of these feeds are either not edible or not very palatable to human beings. Pig is non ruminant/ monogastric animal so cannot utilize fibrous fodder efficiently. Due to its simple stomach, ration must have more concentrate preferably in ground form.

Objectives

1. To study the nutrient requirements of different categories of swine.
2. To formulate diets for different categories of swine.

Nutrient requirements of swine

Nutrient requirements table serve as guides and are used to formulate nutritionally adequate and economically practical diets. The NRC, ARC as well as BIS have formulated feeding standards for pigs. The NRC (1998) nutrients requirements for swine are expressed on 90% DM basis (Table 1) where as BIS (1986) specification are expressed on 100% DM basis (Table 2) for Indian and crossbred pigs. The BIS has recommended as three categories of pig starter (creep feed), pig grower and pig finishing/breeding meal under specification number (IS 7472: 1986; Clause 3.3-3.5) reaffirmed in 2001. ARC standards were given in 1981. Till date these standards are not modified, hence outdated and now an inappropriate guide to the nutrient requirements of swine.

Common feedstuffs for swine

1. Grains and their byproducts are most important sources of energy in swine diets. Most common cereal grains like maize, wheat, barley, sorghum, etc,

are quite palatable and digestible. However, cereal byproducts tend to be more variable; therefore, their use in swine diets may be limited.

Table 1: Nutrient requirement of swine (NRC, 1998 on 90% DM basis)

Intake parameters	Live weight (kg)						Lac-tating	Preg-nant	Boar
	1-5	5-10	10-20	20-50	50-80	80-120			
Protein (%)	26	23.7	20.9	18	15.5	13.2	18.4	12.4	13
ME (kcal/kg)	3265	3265	3265	3265	3265	3265	3265	3265	3265
Lysine (%)	1.34	1.19	1.01	0.83	0.66	0.52	0.85	8.7	0.6
Methionine+Cystine (%)	0.76	0.68	0.58	0.47	0.39	0.31	0.41	6.1	0.42
Threonine (%)	0.84	0.74	0.63	0.52	0.43	0.34	0.52	6.9	0.5
Tryptophan (%)	0.24	0.22	0.18	0.15	0.12	0.10	0.15	1.7	0.12
Phenylalanine + Tyrosine (%)	1.26	1.12	0.95	0.78	0.63	0.49	0.95	8.6	0.57
Linoleic acid (%)	0.1	0.1	0.1	0.1	0.1	0.1	0.1	0.1	0.1
Calcium (%)	0.90	0.80	0.70	0.60	0.50	0.45	0.75	0.75	0.75
Phosphorus (%)	0.70	0.65	0.60	0.50	0.45	0.40	0.60	0.60	0.60
Pantothenic acid (mg/kg)	12	10	9	8	7	7	12	12	12
Riboflavin (mg/kg)	4	3.5	3	2.5	2	2	3.75	3.75	3.75
Vitamin B_{12} (μg/kg)	20	17.5	15	10	5	5	15	15	15

Table 2: BIS specification for pig feed (under specification number IS 7472: 1986, Clause 3.3-3.5 reaffirmed in 2001)

Characteristics	Creep feed	Growth meal	Finishing/Breeding meal
Moisture (%) max.	11	11	11
CP (%) min.	20	18	16
Crude fat (%) min.	2	2	2
Crude fibre (%) max.	5	6	8
Total ash (%) max.	8	8	8
Acid insoluble ash (%) max.	4	4	4
ME (kcal/kg) min.	3360	3170	3170
Calcium (%) min.	0.6	0.6	0.6
Phosphorus (%) min.	0.6	0.4	0.5
Iron (mg/kg) min.	100	90	80
Copper (mg/kg) min.	30	30	20
Zinc (mg/kg) min.	50	50	50
NaCl (%) max.	0.5	0.5	0.5

Characteristics	Creep feed	Growth meal	Finishing/Breeding meal
Niacin (mg/kg)	17	14	10
Pantothenic acid (mg/kg)	11	10	10
Riboflavin (mg/kg)	3	2.4	2.2
Vitamin B_{12} (µg/kg)	15	11	11

2. In addition, fat and oils are often used to increase the energy density of swine diets. Fats and oils such as lard, choice white grease, beef tallow, coconut oil and soybean oil contain about 2.25 times as much ME as most of the cereal grains.
3. Fat additions will usually increase the cost of diet, which must be offset by an increase in pig performance. Research indicates that the addition of 1-5% fat to growing finishing swine diets will improve feed conversion and often average daily gain with no adverse effect on carcass quality. A reduction in the amount of dust will be evident and wear on mixing equipment and augers will be reduced with 2-3% added fat. Addition of fat above 5% will further improve feed conversion, but physical handling problems such as bridging in the feeders and caking in the mixer may limit the use of these higher levels.
4. Diets containing fats may become rancid during prolonged storage are when feed is exposed to high temperature. Therefore, an antioxidant such as ethoxyquin, BHT or BHA should be added to fat before mixing it into the rations.
5. The other sources of energy for swine include roots and tubers, such as cassava, potatoes, sweet potatoes, bananas and molasses. Molasses can be used up to 5% in swine ration as energy source.
6. Protein sources are used to compliment cereal grains to meet the amino acids needs of the pigs. The plant protein supplements used in feeding of pigs are soybean meal, ground nut cake/meal, cotton seed meal, linseed meal, peanut meal, canola meal, safflower meal and sunflower meal.
7. The legume seeds contain 20-30% protein and can also be used for pig feeding after removal of growth inhibitors by heat treatment.
8. Pigs are more prone to gossypol poisoning. Cotton seed meals with high gossypol content are not generally recommended. Ground nut cake is highly palatable to pigs.
9. The animal proteins are generally more expensive than plant proteins, but they usually contain a better balance of amino acids of greater bioavailability and are good source of vitamins and trace mineral elements. The common animal protein supplements include fish meal, meat meal, blood meal, milk whey, etc.

10. The pigs have a simple digestive tract and in general they prefer coarsely ground material instead of fine mixture. Pigs have limited ability to digest fibers (only 5-8%).
11. The full grown pig can conveniently consume 3-5 kg succulent green fodder which is also helpful for economical pig forming.
12. The vegetables like pumpkin, carrots etc are kitchen waste can also be used for feeding of pigs.

Table 3: Maximum percentage level of incorporation of feedstuffs in ration of swine

Ingredients	Type of pigs			
	Gestation	Lactation	Starter	Grower/ Finisher
Alfalfa meal (dehydrated)	90	10	0	5
Animal fats stabilized	5	5	5	10
Bakery surplus material	40	40	20	40
Blood meal	3	3	-	3
Barley	80	80	25	85
Bajra	80	80	60	85
Cotton seed meal	5	5	0	5
Fish meal	10	10	15	10
Ground nut cake	20	20	20	20
Gaur meal toasted	10	10	-	15
Hominy feed	20	20	-	20
Linseed meal	10	10	5	10
Maize	80	80	60	85
Maize gluten	8	5	3	8
Meat meal	10	5	-	10
Meat and bone meal	10	10	5	5
Molasses	5	5	5	5
Mustard cake (Deoiled)	5	5	5	5
Oats	70	30	20	50
Potatoes	80	-	-	30
Rice kani	40	25	10	45
Rice bran	50	50	-	50-60
Sorghum	85	85	60	80-90
Skim milk powder	-	-	40	-
Sugars	-	-	5	-
Soybean	30	25	30	25
Soybean meal (full fat)	20	20	30	25
Sun flower meal	10	5	-	10

Ingredients	Type of pigs			
	Gestation	Lactation	Starter	Grower/ Finisher
Triticale	80	80	20	85
Til cake	15	15	10	15
Wheat	80	80	60	85
Wheat bran	30	10	-	20
Whey	5	5	20	10
Yeast	3	3	3	3

Formulation of swine diet

The principles of formulation of swine diets are to supply required nutrients through a feed mix of the available feed ingredients at low cost to achieve optimum growth, production of litter and feed utilization. From nutritional stand point, there is no single best formula in terms of ingredients that are used. Ingredients, therefore, selected on the basis of availability, price and quality of nutrients they contain but they must be sound, palatable and free from any adulteration. A large number of agricultural and industrial by products available, may not contain optimum concentration of nutrients but contain some deleterious factors, are also included in livestock feed within their safe and/ are effect level of inclusion to economize livestock production without affecting health. Before computing a balanced ration for the pigs, one should consider breed (meat or lard type), physiological stage (suckling piglets, weaners or growing, fattening, gestation, lactation and breeder boar), nutritional requirements (NRC/BIS specifications for different categories), availability and chemical composition of feed stuffs, maximum percent level of incorporation of feed ingredients, anti quality factors if any and cost of feed ingredients.

Methods of computing the diets

There are four method for computing the diets i.e. Hit and trial method, Pearson square method, algebraic equation method and linear programming.

Pearson's square method

The rations can easily be computed by using the following steps for a particular animal for which feed is to be formulated.

a. Draw a square with diagonals.

b. Write the desired protein percentage in the center where the diagonals intercept each other.

c. Write the percent of protein contents of the basal mixture and protein mixtures to be mixed on the left side corners of the square.

d. Take out differences diagonally between the figures on the left hand side and the center figure (desired protein content). Write the differences on right hand corners of the square in the direction of diagonal lines.

e. The figures obtained on the right side coroners of the parts or proportion in which ratio the given feeds should be mixed to obtain desired protein percentage in the resultant mixture.

Examples of diet formulation:

Exercise: Formulate a complete mixed diet for a pig weighing 40 kg with the ingredients given below. The diets should provide 16.0% protein, 0.65% calcium, 0.5% phosphorus and 0.74 % lysine. A trace mineral TM salt, vitamin supplement and common salt are added at levels of 1.0, 0.1, and 0.5 % respectively.

Ingredients	Nutrient content						
	DE (Kcal/ kg)	Protein (%)	Ca (%)	P (%)	Lysine (%)	TM salt (%)	Vitamin Mix/ Common salt
Maize	3.870	9.0	0.02	0.26	0.24	-	-
Lucerne hay	2.460	17.0	0.25	0.64	0.80	-	-
Groundnut cake	-	42.0	0.17	0.57	1.40	-	-
Fish meal	3.650	60.0	5.00	3.30	5.44	-	-
Lime stone	3.260	-	34.0	-	-	-	-
Dical PO_4	-	-	28.0	17.00	-	-	-
Soybean meal	-	45.0	0.25	0.64	2.86	-	-
TM salt	-	-	-	-	-	100.00	-
Vitamin Mix.	-	-	-	-	-	-	100.00
Common salt	-	-	-	-	-	-	100.00

DIET A

Step 1: Calculate the sum of ingredients that do not contribute protein to the diet.

Ingredient	%
Ca- P supplement	1.5
TM salt	1.0
Common salt	0.5
Vitamin mix.	0.1
Total	3.1

Step II: Calculate the percent protein needed from 96.9% (100-3.1) of the diet

$(16/96.9) \times 100 = 16.50$ %

Step III: The essential amino acids content of the available ingredients should be kept in consideration to calculate the CP % supplied by the individual ingredients so that the amino acid requirement of the animal can be satisfied.

Diet A	Total feed %	Total Lysine (%)
Maize	72	0.17
Lucerne hay	8	0.06
Groundnut cake	14	0.20
Fish meal	6	0.32
Total		0.75

Maize 9.00 X 0.9 = 8.1
L.hay 17.0 X 0.1 = 1.7
= 9.8

GNC 42.0 X 0.7 = 29.4
PM 62.0 X 0.3 = 18.0
= 47.4

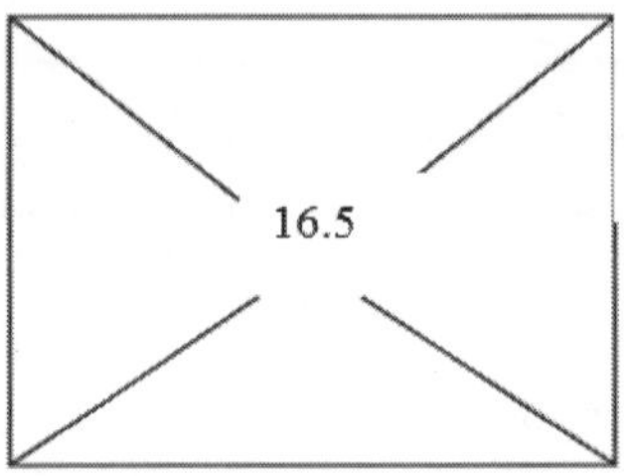

30.9 (30.9/37.6) x 96.9 = 9.66
Maize = 71.70
L.hay = 07.96

6.7 (6.7/37.6) x 96.9 = 17.24
GNC = 12.07
FM = 5.17

Thus assuming that diet contains 80% energy feeds and 20% protein feeds, if 10% maize is replaced by Lucerne meal and protein feeds (groundnut cake and fish meal are in proposition of 70: 30), the lysine requirement of the animal is met. Similarly, requirement of other essential amino acids can be worked out, if specified.

Step IV: Calculate percent ingredients added to give 16.5 % protein from 96.9 % total diet.

Step V: Calculate calcium and phosphorus contribution from known ingredients. Subtract these from the requirements to obtain the deficit quantity. Then add Ca-P supplements (Limestone and dicalcium phosphate) to meet the dietary requirements.

Attributes		**Requirements***	
		Calcium (0.65 kg)	**Phosphorus (0.50 kg)**
Ingredients	**Parts (%)**		
Maize	71.70	0.014	0.186
LM	7.96	0.020	0.052
GNC	12.07	0.020	0.069
Fish meal	5.17	0.258	0.171
Total		0.312	0.478
Deficit		0.338	0.022

*Ca & P supplied by given parts

Step VI: Final diet:

Ingredients	Diet		Supplements	
	Computed	Adjusted	Computed	Adjusted
Maize	71.70	72.20	-	-
Lucerne hay	7.96	7.96	28.66	28.58
Groundnut cake	12.07	12.07	43.45	43.45
Fish meal	5.17	5.17	18.81	18.61
TM salt	1.00	1.00	3.60	3.60
Lime stone	1.00	1.00	3.60	3.60
Vitamin Mix.	0.10	0.10	0.36	0.36
Common salt	0.50	0.50	1.80	1.80
Total	99.50	100.00	100.08	100.00

Calcium = (0.338/0.340) ×100 = 1.00 % limestone (Approx.)

- Maize can be used to make the total 100%.
- To convert non-maize ingredients into a fortified supplement that can be premixed or fed free choice with maize multiply non-maize ingredients by : 100/100 – 72.2 = 3.60
- Groundnut cake is used as a filler or adjuster to make the supplements to 100 %

DIET B: Using maize and soybean meal (SBM) as energy and protein sources, respectively.

Step 1:

Ingredient	%
Ca- P supplement	2.0
TM salt	1.0
Common salt	0.5
Vitamin mix.	0.1
Total	3.6

STEP II: Percent protein needed from 96.4% of the diet (16/96.4) ×100 = 16.6%

STEP III: Using maize and SBM in the ratio of 80:20, calculate lysine contribution

Ingredient	Total feed (%)	Total lysine (%)
Maize	80	0.192
SBM	20	0.572
Total	100	0.764

Thus it meets the lysine requirements of the pig.

Step IV: Calculate Percent ingredients added to give 16.6% protein from 96.4 % total diet.

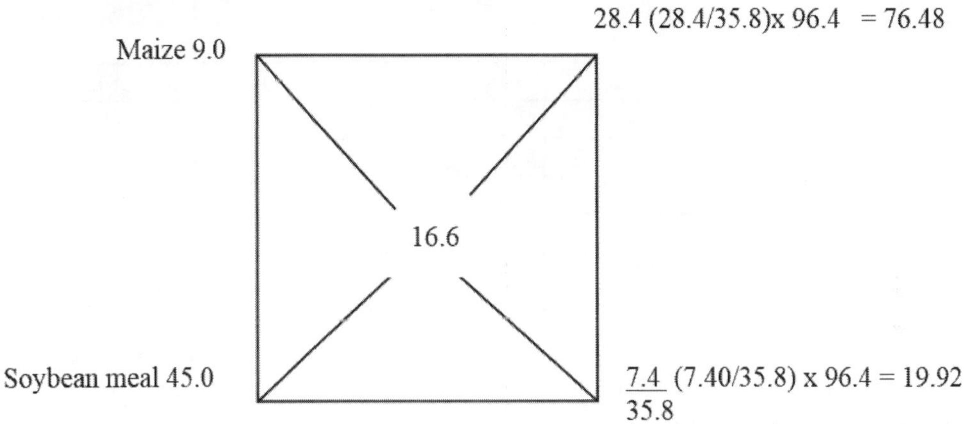

Step V: Calculate Ca and P contribution of known ingredients. Subtract these from the requirements to obtain deficit quantity and then add Ca-p supplement to meet the requirements.

Attributes			
		Calcium (0.65 kg)	**Phosphorus (0.50 kg)**
Ingredients	**Parts (%)**		
Maize	76.48	0.015	0.199
SBM	19.92	0.049	0.129
Total		0.064	0.328
Deficit		0.586	0.172

*Ca & P supplied by given parts

Phosphorus content from dicalcium phosphate (DCP) = (0.172/0.17) × 100 = 1.00 kg

1.0 kg DCP provides 23 % calcium, so, deficiency of calcium = 0.586 – 0.230 = 0.356 kg

This deficiency can be makeup by adding limestone, which contains 34 % calcium.

Quantity of limestone to be added to makeup the deficiency: (0.356/0.34) × 100 = 1.04 %

Step VI: Final diet:

Ingredient	Diet		Supplements	
	Computed	Adjusted	Computed	Adjusted
Maize	76.44	76.44	-	-
SBM	19.92	19.92	84.46	84.57
TM Salt	1.00	1.00	4.24	4.24
Lime stone	1.04	1.04	4.41	4.41
Dicalcium PO_4	1.00	1.00	4.24	4.24
Common salt	0.50	0.50	2.12	2.12
Vitamin Mix.	0.10	0.10	0.42	0.42
Total	100.00	100.00	99.89	99.89

To convert non- maize ingredients into a fortified supplement multiply the non- grain ingredients by 100/100- 76.44 = 4.24. SBM is used as a filler or adjustor to make the supplements to 100%.

CHAPTER 38

NUTRIENT REQUIREMENTS AND FEED FORMULATION FOR HORSES

Traditionally horse feeding practices in India involves the use of crushed oat, wheat bran, gram, jaggery, Lucerne and grass hay. Race horses were traditionally fed large quantities of oats as their principal energy source. Nutritional needs of horses vary greatly. Lot depends on the type of horse, its age, and weight and on the work it does.

Objectives:

1. To calculate the nutrient requirements of different categories of horses.
2. To formulate rations for different categories of horses.

Procedure:

Horses can consume approximately 2.0 to 2.5 percent of their body weight on DM basis each day, so a horse weighing 500 kg can eat 10 to 12.5 kg of food. Foals less than six months of age can eat 2.0 to 4.0 percent of their body weight per day. 50 percent or more of the horse diet by weight should be from forages. If a horse is working hard and requires more energy the use of grain is increased and percentage of forages can be reduced. However, forage amount should never go below 1.0 percent of body weight per day. Following table 1 can be used as guide line of feed and roughage required for different categories of horses.

Table 1: Feed required in kg by a horse per 100 kg live weight (DM basis)

Category	Total Feed	Roughage	Concentrate
Mature horse, idle	1.5	1.5	-
Light work (Less than 2 hrs/ day)	1.75-2.25	1.25-1.5	0.5-0.75
Medium work (2-4 hrs /day)	2.25-2.75	1.25-1.75	1.0
Heavy work (over 4 hrs /day)	2.0-2.5	1.0	1.0-1.5
Pregnant / Lactating mare	1.5-3.0	0.75-1.5	0.75-1.5

Different categories of horses require nutrient concentration in their total diets presented in Table 2.

Table 2: Minimum nutrient concentration needed in the diets of different categories of Horses

Category	DE Mcal / kg	%			Vit. A IU/ kg
		CP	Calcium	Phosphorus	
Idle horse at maintenance	2.00	8.0	0.3	0.2	1830
Breeding season (light work)	2.40	10.0	0.35	0.25	2640
Breeding season (moderate work)	2.65	10.5	0.35	0.25	2640
Pregnant Mare (09 months)	2.25	10.0	0.43	0.32	3710
Pregnant mare (10 months)	2.25	10.0	0.43	0.32	3650
Pregnant mare (11 months)	2.40	10.6	0.45	0.34	3650
Lactating mare (foaling – 3 months)	2.60	13.2	0.52	0.34	2750
Lactating mare (3 months – weaning)	2.45	11.0	0.36	0.22	3020
Working horse (light work)	2.45	10.0	0.30	0.22	2690
Working horse (moderate Work)	2.65	10.5	0.31	0.23	2420
Working horse (intense work)	2.85	11.5	0.35	0.25	1950
Growing horse (weaning 4 months)	2.90	14.5	0.68	0.38	720
Growing horse (weaning -6 months- moderate growth)	2.90	14.5	0.56	0.31	850
Growing horse (weaning -6 months –rapid growth)	2.90	14.5	0.61	0.34	740
Yearling, 12 months (moderate Growth)	2.80	12.6	0.43	0.24	980
Yearling, 12 months (rapid growth)	2.80	12.6	0.45	0.25	870
Two year old horse (not in training)	2.45	10.4	0.31	0.17	1200
Two year old horse (in training)	2.65	11.3	0.34	0.20	930

Table 3: Nutritive value of common roughages (on as fed basis).

Roughage	DM	DCP	TDN	Ca	P
Legume hay	88	10	60	1.5	0.25
Legume fodder	15	3.0	12	0.3-0.5	0.05-0.10
Green grasses	25	1.5	12	-	-
Cereal fodders	20	1.0	13	0.08	0.045
Cereal hay	88	3.0	51	0.3	0.2

Table 4: Daily energy and protein requirement for different categories of horses

Type of Horse	Mature Wt. (kg)	Daily live wt. gain (Kg/day)	Digestible energy (MJ)	Crude protein (g)
Weanling (4 months)	400	0.85	56.5	675
	500	0.85	60.2	720
	600	1.00	69.0	825
Weanling (6 months; moderate growth)	400	0.55	54.0	643
	500	0.65	62.8	750
	600	0.75	71.1	850
Weanling (6 months; Rapid growth)	400	0.70	60.7	725
	500	0.85	72.0	860
	600	0.95	80.3	960
Yearling (12 months; moderate growth)	400	0.40	65.3	700
	500	0.50	71.9	851
	600	0.65	95.0	1023
Yearling (12 months; rapid growth)	400	0.50	71.5	770
	500	0.65	89.1	956
	600	0.80	105.0	1127
Two year old	400	0.15	64.0	650
	500	0.20	78.7	800
	600	0.30	98.3	998
Maintenance	400	-	56.1	563
	500	-	68.6	656
	600	-	81.2	766
Pregnant (9 month)	400	-	62.3	654
	500	-	76.1	801
	600	-	90.0	947
Pregnant (10 months)	400	-	63.2	666
	500	-	77.4	815
	600	-	91.6	965

Type of Horse	Mature Wt. (kg)	Daily live wt. gain (Kg/day)	Digestible energy (MJ)	Crude protein (g)
Pregnant (11 months)	400	-	67.4	708
	500	-	82.4	866
	600	-	97.5	1024
Lactating (foaling to 3 months)	400	-	95.8	1141
	500	-	118.4	1427
	600	-	141.0	1711
Lactating (3 month to weaning)	400	-	82.4	839
	500	-	101.7	1048
	600	-	120.9	1258

Steps to calculate daily feed requirement for horses of different categories.

Step 1

Estimate the weight of each horse

Step 2

Determine the nutritional needs of each horse

Steps 3

Ascertain the nutritive value of common feed.

Step 4

Determine how much feed each horse can consume daily. This enables us to calculate if certain feeds can be consumed in sufficient quantities to provide the require nutrition.

Step 5 Calculate the feed required for each horse.

CHAPTER 39

FORMULATION OF BALANCE DIETS FOR DOGS AND CATS

Dogs (*Canis familiaris*) and cats (*Felis catus*) are in the order Carnivora. However, feeding behavior-wise dogs are best described as omnivorous animals, while cats are true carnivores. Numerous factors affect the nutritional requirements of dogs and cats. These include age, breed, physiological growth status, sex, activity, health, reproduction and environmental conditions. The National Research Council (NRC), Association of American Feed Control Officials (AAFCO) and European Pet Food Industry Federation (FEDIAF) recommended the nutrient values based on scientific principles and take into account the requirements for practical feeding of dog and cats.

Objectives:

1. To calculate the nutrient requirements of different categories of dogs and cats.
2. To formulate balanced diets for different categories of dogs and cats.

NRC (2006) requirements are based on the following points

1. The energy density of diet for both dogs and cats was assumed to be 4,000 kcal ME per kg.
2. Requirements for growth of puppies are based on a 5.5 kg puppy that consumes 1,000 kcal ME per day.
3. Requirements for adult dogs at maintenance are based on a 15 kg adult dog that consumes 1,000 kcal ME per day.
4. Requirements of gestating and lactating dogs are based on a 22 kg bitch with eight puppies in peak lactation consuming 5,000 kcal ME per day.
5. Requirements for growth of kittens are based on an 800 g kitten that consumes 180 kcal ME per day.

6. Requirements for adult cats at maintenance are based on a 4 kg adult cat that consumes 250 kcal ME per day.
7. Requirements of gestating and lactating cats are based on a 4 kg queen with four kittens in peak lactation consuming 540 kcal ME per day.

NRC requirements are given as minimal requirements (MR), adequate intake (AI), recommended allowance (RA) and safe upper limit (SUL) as an example of how NRC (2006) requirements for dogs and cats have been presented in tables later.

Minimal requirement (MR): It is defined as the minimal concentration or amount of a bioavailable nutrient that will support a defined physiological state.

Adequate intake (AI): It is defined as the concentration in the diet or amount of nutrient required by the animal that is presumed to sustain a given life stage when no minimal requirement has been demonstrated.

Recommended allowance (RA): It is based on the minimal requirement and may include a bioavailable factor. Where minimal requirement is not available, the RA is based on the AI.

Safe upper limit (SUL): It is the maximal concentration or amount of a nutrient that has not been associated with adverse effects.

Nutrients requirement of dogs and cats as per different feeding standard

Nutrient requirement of dogs

Table 1. The nutrient requirements (NRC, 2006) are given here in nutshell for easy application.

ME requirements per day per kg $W^{0.75}$	
Maintenance	130 kcal
Growing	260 kcal
Adolescence	200 kcal
Pregnancy	186 kcal
Lactation	460 kcal
Fat requirements per day per kg $W^{0.75}$	
Maintenance	1.8 g
Growth	up to 5.9 g
Protein requirements per day per kg $W^{0.75}$	
Maintenance	3.28 g
Growth	up to 15.7 g

Table 2. Nutrient requirement for growth of puppies after weaning (NRC, 2006)

Nutrient	Minimum Recommended Nutrient Levels for Dogs – Unit per 100 g dry matter			Recommended Allowance (RA)			Safe Upper Limit (SUL)		
	Amt./ kg DM (4000 kcal)[a]	Amt./ 1000 kcal ME[b]	Amt./ kg BW$^{0.75}$[c]	Amt./ kg DM (4000 kcal)[a]	Amt./ 1000 kcal ME[b]	Amt./ kg BW$^{0.75}$[c]	Amt./ kg DM (4000 kcal)[a]	Amt./ 1000 kcal ME[b]	Amt./ kg BW$^{0.75}$[c]
Growing puppies 4-14 weeks old									
Crude Protein* (g)	**180**	**45**	**12.5**	**225**	**56.3**	**15.7**			
Amino Acids									
Arginine (g)	6.3	1.58	0.44	7.9	1.98	0.55			
Histidine (g)	3.1	0.78	0.22	3.9	0.98	0.27			
Isoleucine (g)	5.2	1.30	0.36	6.5	1.63	0.45			
Leucine (g)	10.3	2.58	0.19	12.9	3.22	0.90			
Lysine* (g)	7.0	1.75	0.39	8.8	2.20	0.61	>20	>5.0	>1.39
Methionine* (g)	2.8	0.70	0.72	3.5	0.88	0.24			
Methionine + cysteine* (g)	5.6	1.40	0.49	7.0	1.75	0.49			
Phenylalanine (g)	5.2	1.30	0.36	6.5	1.63	0.45			
Phenylalanine + tyrosine* (g)	10.4	2.60	0.72	13.0	3.25	0.90			
Threonine (g)	6.5	1.63	0.45	8.1	2.03	0.56			
Tryptophan (g)	1.8	0.45	0.13	2.3	0.58	0.16			
Valine (g)	5.4	1.35	0.38	6.8	1.70	0.47			
	Growing puppies 14 weeks and older								
Crude Protein* (g)	**140**	**35**	**9.7**	**175**	**43.8**	**12.2**			
Amino Acids									
Arginine (g)	5.3	1.33	0.37	6.6	1.65	0.46			
Histidine (g)	2.0	0.50	0.14	2.5	0.63	0.17			
Isoleucine (g)	4.0	1.00	0.28	5.0	1.25	0.35			
Leucine (g)	6.5	1.63	0.45	8.2	2.05	0.57			

Nutrient	Minimum Recommended Nutrient Levels for Dogs – Unit per 100 g dry matter			Recommended Allowance (RA)			Safe Upper Limit (SUL)		
	Amt./ kg DM (4000 kcal)[a]	Amt./ 1000 kcal ME[b]	Amt./ kg $BW^{0.75c}$	Amt./ kg DM (4000 kcal)[a]	Amt./ 1000 kcal ME[b]	Amt./ kg BW $^{0.75c}$	Amt./ kg DM (4000 kcal)[a]	Amt./ 1000 kcal ME[b]	Amt./ kg BW $^{0.75c}$
Lysine* (g)	5.6	1.40	0.39	7.0	1.75	0.49	>20	>5.0	>1.39
Methionine* (g)	2.1	0.53	0.15	2.6	0.65	0.18			
Methionine + cysteine* (g)	4.2	1.05	0.29	5.3	1.33	0.37			
Phenylalanine (g)	4.0	1.00	0.28	5.0	1.25	0.35			
Phenylalanine + tyrosine* (g)	8.0	2.0	0.56	10.0	2.50	0.70			
Threonine (g)	5.0	1.25	0.35	6.3	1.58	0.44			
Tryptophan (g)	1.4	0.35	0.10	1.8	0.45	0.13			
Valine (g)	4.5	1.13	0.31	5.6	1.40	0.39			
Total Fat (g)				**85**	**21.3**	**5.9**	**330**[a]	**82.5**	**23.0**
Linoleic acid (ω-6) (g)				13	3.3	0.8	65[a]	16.3	4.5
Alpha-linolenic (ω-3) (g)				0.8	0.2	0.05			
Arachidonic acid (g)				0.3	0.08	0.022			
E Eicosapentaenoic + Docosahexaenoic acid (ω-3) (g)				0.5	0.13	0.036	11[a]	2.8	0.77
Minerals									
Calcium (g)	8.0	2.0	0.56	12	3.0	0.68	18	4.5	1.25
Phosphorus (g)				10	2.5	0.68			
Magnesium (mg)				400	100	27.4			
Sodium (mg)				2200	550	100			
Potassium (g)				4.4	1.1	0.30			

Nutrient	Minimum Recommended Nutrient Levels for Dogs – Unit per 100 g dry matter			Recommended Allowance (RA)			Safe Upper Limit (SUL)		
	Amt./ kg DM (4000 kcal)[a]	Amt./ 1000 kcal ME[b]	Amt./ kg $BW^{0.75c}$	Amt./ kg DM (4000 kcal)[a]	Amt./ 1000 kcal ME[b]	Amt./ kg BW $^{0.75c}$	Amt./ kg DM (4000 kcal)[a]	Amt./ 1000 kcal ME[b]	Amt./ kg BW $^{0.75c}$
Chloride (mg)				2900	720	200			
Trace elements									
Iron (mg)	72	18	5.0	88	22	6.1			
Copper (mg)				11	2.7	0.76			
Zinc (mg)	40	10	2.7	100	25	6.84			
Manganese (mg)				5.6	1.4	0.38			
Selenium (µg)	210	52.5	13.7	350	87.5	25.1			
Iodine (µg)				880	220	61.0			
Vitamins									
Vitamin A (RE)				1515	379	105	15000	3750	1044
Vitamin D (µg)				13.8	3.4	0.96	80	20	5.6
Vitamin E (mg)				30	7.5	2.1			
Thiamine (mg)				1.38	0.34	0.096			
Riboflavin (mg)				5.25	1.32	0.37			
Pantothenic acid (mg)				15.0	3.75	1.04			
Vitamin B6 (Pyridoxine) (mg)				1.5	0.375	0.10			
Vitamin B_{12} (µg)				35	8.75	2.4			
Folic Acid (µg)				270	68	18.8			
Choline (mg)				1700	425	118			
Vitamin K (mg)				1.64	0.41	0.11			

Table 3. Minimum recommended nutrient levels for Dogs– Unit per 100 g DM (FEDIAF, 2011)

Nutrient	Minimum Recommended Nutrient Levels for Dogs – Unit per 100 g DM				
	Unit	Adult	Early Growth (< 14 weeks) & Reproduction	Late Growth (14 weeks)	Maximum (L) = legal (N) = nutritional
Minimum recommended					
Protein*	g	**18.0**	**25.0**	**20.0**	-
Arginine	g	0.52	0.82	0.69	-
Histidine	g	0.23	0.39	0.25	-
Isoleucine	g	0.46	0.65	0.50	-
Leucine	g	0.82	1.29	0.80	-
Lysine*	g	0.42	0.88	0.70	Growth: 2.8 (N)
Methionine*	g	0.31	0.35	0.26	-
Methionine + cysteine*	g	0.62	0.70	0.53	-
Phenylalanine	g	0.54	0.65	0.50	-
Phenylalanine + tyrosine*	g	0.89	1.30	1.00	-
Threonine	g	0.52	0.81	0.64	-
Tryptophan	g	0.17	0.23	0.21	-
Valine	g	0.59	0.68	0.56	-
Fat*	g	5.5	8.50	8.50	-
Linoleic acid (ω-6) *	g	1.32	1.30	1.30	Early growth: 6.50 (N)
Arachidonic acid (ω-6)	mg	-	30.0	30.0	
Alpha-linolenic acid (ω-3) *	g	-	0.08	0.08	
EPA + DHA (ω-3) *	g	-	0.05	0.05	
Minerals					
Calcium	g	0.50	1.00	0.80[a] - 1.00[b]	Adult: 2.5 (N) Early growth: 1.6 (N) Late growth 1.8 (N)

Nutrient	Minimum Recommended Nutrient Levels for Dogs – Unit per 100 g DM				
	Unit	Adult	Early Growth (< 14 weeks) & Reproduction	Late Growth (14 weeks)	Maximum (L) = legal (N) = nutritional
Phosphorus	g	0.40	0.90	0.70	Adult: 1.60 (N)
Ca/P ratio	g	1/1-2/1	1/1-1.6/1	1/1-1.6/1[b] or1.8/1[a]	-
Potassium	g	0.50	0.44	0.44	-
Sodium	g	0.10	0.22	0.22	Adult: 1.80 (N)
Chloride	g	0.15	0.33	0.33	Adult: 2.25 (N)
Magnesium	g	0.07	0.04	0.04	-
Trace elements					
Copper	mg	0.72	1.10	1.10	2.8 (L)
Iodine	mg	0.11	0.15	0.15	1.1 (L)
Iron	mg	3.60	8.80	8.80	142 (L)
Manganese	mg	0.58	0.56	0.56	17.0 (L)
Selenium	μg	30.0	35.0	35.0	56.8 (L)
Zinc	mg	7.2	10.0	10.0	28.4 (L) Growth: 100 (N)
Vitamins					
Vitamin A	IU	500	500	500	40000 (N)
Vitamin D	IU	50.0	55.2	50.0	227 (L) 320 (N)
Vitamin E	IU	3.60	5.00	5.00	-
Thiamine	mg	0.23	0.14	0.14	-
Riboflavin	mg	0.60	0.53	0.53	-
Pantothenic acid	mg	1.00	1.50	1.50	-
Vitamin B_6 (Pyridoxine)	mg	0.15	0.15	0.15	-
Vitamin B_{12}	μg	2.20	3.50	3.50	-
Niacin	mg	1.10	1.70	1.70	-
Folic Acid	mg	18.0	27.0	27.0	-
Biotin	μg	-	-	-	-
Choline	mg	120	170	170	-

Table 4. AAFCO dog food nutrient profiles based on DM

Nutrient	Minimum Recommended Nutrient Levels for dogs – Unit per 100 g dry matter			
	Unit	Growth and Reproduction Minimum	Adult Maintenance Minimum	Maximum
Minimum recommended				
Protein*	%	**22.5**	**18.0**	
Arginine	%	1.00	0.51	
Histidine	%	0.44	0.19	
Isoleucine	%	0.71	0.38	
Leucine	%	1.29	0.68	
Lysine*	%	0.90	0.63	
Methionine*	%	0.35	0.33	
Methionine + cysteine*	%	0.70	0.65	
Phenylalanine	%	0.83	0.45	
Phenylalanine + tyrosine*	%	1.30	0.74	
Threonine	%	1.04	0.48	
Tryptophan	%	0.20	0.16	
Valine	%	0.68	0.49	
Crude Fat*	%	**8.5**	**5.5**	
Linoleic acid (ω-6)	%	1.3	1.1	
Alpha-linolenic acid (ω-3)	%	0.08	ND	
E Eicosapentaenoic + Docosahexaenoic acid (ω-3)	%	0.05	ND	
(Linoleic + Arachidonic):(alpha Linolenic + Eicosapentaenoic + Docosahexaenoic) acid Ratio				30:1
Minerals				
Calcium	%	1.2	0.5	2.5 (1.8)
Phosphorus	%	1.0	0.4	1.6

Nutrient	Minimum Recommended Nutrient Levels for dogs – Unit per 100 g dry matter			
	Unit	Growth and Reproduction Minimum	Adult Maintenance Minimum	Maximum
Ca/P ratio	%	1:1	1:1	2:1
Potassium	%	0.60	0.6	
Sodium	%	0.30	0.08	
Chloride	%	0.45	0.12	
Magnesium	%	0.06	0.06	
Trace elements				
Copper	mg/kg	12.4	7.3	
Iodine	mg/kg	1.0	1.0	11
Iron	mg/kg	88	40	
Manganese	mg/kg	7.2	5.0	
Selenium	mg/kg	0.35	0.35	2
Zinc	mg/kg	100	80	
Vitamins				
Vitamin A	IU/kg	5000	5000	250000
Vitamin D	IU/kg	500	500	3000
Vitamin E	IU/kg	50	50	-
Thiamine	mg/kg	2.25	2.25	-
Riboflavin	mg/kg	5.2	5.2	-
Pantothenic acid	mg/kg	12	12	-
Vitamin B6 (Pyridoxine)	mg/kg	1.5	1.5	-
Vitamin B_{12}	mg/kg	0.028	0.028	-
Niacin	mg/kg	13.6	13.6	-
Folic Acid	mg/kg	0.216	0.216	-
Choline	mg/kg	1360	1360	-

[a]Presumes a caloric density of 4000 kcal ME/kg, as determined in accordance with Regulation PF9.

Nutrient requirements of Cats

Owing to the small variation in adult body weights, the energy needs of cats can be expressed per kg BW instead of per kg metabolic weight. In addition, if metabolic weight is used to calculate MER with the intra-specific allometric coefficient of 0.67

proposed by Heusner in 1991 should be used (NRC, 2006a), which has recently been confirmed to be a more accurate than the 0.75 (Nguyen *et al.* 2001; Edtstadtler-Pietsch 2003).

Table 5. Nutrient requirement of Cats (FEDIAF, 2011)

Nutrient	**Minimum Recommended Nutrient Levels for Cats – Unit per 100 g DM**			
	Unit	**Adult**	**Growth and / Reproduction***	**Maximum (L) = legal (N) = nutritional**
Minimum recommended				
Protein*	**g**	**25.0**	**28.0 / 30.0**	**-**
Arginine	g	1.00	1.07 / 1.11	Growth: 3.5 (N)
Histidine	g	0.30	0.33	-
Isoleucine	g	0.49	0.54	-
Leucine	g	1.17	1.28	-
Lysine*	g	0.34	0.85	-
Methionine*	g	0.17	0.44	Growth: 1.3 (N)
Methionine + cysteine*	g	0.34	0.88	-
Phenylalanine	g	0.46	0.50	-
Phenylalanine + tyrosine*	g	1.76	1.91	-
Threonine	g	0.60	0.65	-
Tryptophan	g	0.15	0.16	Growth: 1.7 (N)
Valine	g	0.59	0.64	-
Taurine (canned pet food)*	g	0.20	0.25	
Taurine (dry pet food)*	g	0.10	0.10	
Fat*	**g**	**9.0**	**9.0**	**-**
Linoleic acid (ω-6)	g	0.50	0.55	Growth: 6.50 (N)
Arachidonic acid (ω-6)	mg	6.00	20.0	
Alpha-linolenic acid (ω-3)	g	-	0.02	
EPA + DHA (ω-3) *	g	-	0.01	
Minerals				
Calcium	g	0.50	1.00	Adult: 2.5 (N) Early growth: 1.6 (N) Late growth 1.8 (N)

Nutrient	Minimum Recommended Nutrient Levels for Cats – Unit per 100 g DM			
	Unit	Adult	Growth and / Reproduction*	Maximum (L) = legal (N) = nutritional
Phosphorus	g	0.40	0.90	Adult: 1.60 (N)
Ca/P ratio	g	1/1-2/1	1/1-1.6/1	-
Potassium	g	0.50	0.44	-
Sodium	g	0.10	0.22	Adult: 1.80 (N)
Chloride	g	0.15	0.33	Adult: 2.25 (N)
Magnesium	g	0.07	0.04	-
Trace elements				
Copper	mg	0.72	1.10	2.8 (L)
Iodine	mg	0.11	0.15	1.1 (L)
Iron	mg	3.60	8.80	142 (L)
Manganese	mg	0.58	0.56	17.0 (L)
Selenium	µg	30.0	35.0	56.8 (L)
Zinc	mg	7.2	10.0	28.4 (L) Growth: 100 (N)
Vitamins				
Vitamin A	IU	500	500	40000 (N)
Vitamin D	IU	50.0	55.2	227 (L) 320 (N)
Vitamin E	IU	3.60	5.00	-
Thiamine	mg	0.23	0.14	-
Riboflavin	mg	0.60	0.53	-
Pantothenic acid	mg	1.00	1.50	-
Vitamin B6 (Pyridoxine)	mg	0.15	0.15	-
Vitamin B12	µg	2.20	3.50	-
Niacin	mg	1.10	1.70	-
Folic Acid	mg	18.0	27.0	-
Biotin	µg	-	-	-
Choline	mg	120	170	

Table 6. AAFCO cat food nutrient profiles based on DM

Nutrient	Minimum Recommended Nutrient Levels for Cats – Unit per 100 g DM			
	Units DM basis	Growth and Reproduction Minimum	Adult Maintenance Minimum	Maximum
Minimum recommended				
Protein*	%	**30.0**	**26.0**	
Arginine	%	1.24	1.04	
Histidine	%	0.33	0.31	
Isoleucine	%	0.56	0.52	
Leucine	%	1.28	1.24	
Lysine*	%	1.20	0.83	
Methionine*	%	0.62	0.20	1.5
Methionine + cysteine*	%	1.10	0.40	
Phenylalanine	%	0.52	0.42	
Phenylalanine + tyrosine*	%	1.92	1.53	
Threonine	%	0.73	0.73	
Tryptophan	%	0.25	0.16	1.7
Valine	%	0.64	0.62	
Taurine (extruded)	%	0.10	0.10	
Taurine (canned)	%	0.20	0.20	
Crude Fat*	%	**9.0**	**9.0**	
Linoleic acid (ω-6)	%	0.6	0.6	
Alpha-linolenic (ω-3)	%	0.02	ND	
Arachidonic acid	%	0.02	0.02	
E Eicosapentaenoic + Docosahexaenoic acid (ω-3)	%	0.012	ND	
Minerals				
Calcium	%	1.0	0.6	
Phosphorus	%	0.8	0.5	
Potassium	%	0.6	0.6	
Sodium	%	0.2	0.2	
Chloride	%	0.3	0.3	
Magnesium	%	0.08	0.04	

Nutrient	Minimum Recommended Nutrient Levels for Cats – Unit per 100 g DM			
	Units DM basis	Growth and Reproduction Minimum	Adult Maintenance Minimum	Maximum
Trace elements				
Copper (extruded)	mg/kg	15	5	
Copper (canned)	mg/kg	8.4	5	
Iodine	mg/kg	1.8	0.6	
Iron	mg/kg	80	80	
Manganese	mg/kg	7.6	7.6	
Selenium	mg/kg	0.3	0.3	
Zinc	mg/kg	75	75	
Vitamins				
Vitamin A	IU/kg	6668	3332	333300
Vitamin D	IU/kg	280	280	30080
Vitamin E	IU/kg	40	40	-
Vitamin K	mg/kg	0.1	0.1	
Thiamine	mg/kg	5.6	5.6	-
Riboflavin	mg/kg	4.0	4.0	-
Pantothenic acid	mg/kg	5.75	5.75	-
Niacin	mg/kg	60	60	
Vitamin B6 (Pyridoxine)	mg/kg	4.0	4.0	-
Vitamin B_{12}	mg/kg	0.020	0.020	-
Biotin	mg/kg	0.07	0.07	-
Folic Acid	mg/kg	0.8	0.8	-
Choline	mg/kg	2400	2400	-

[a]Presumes a caloric density of 4000 kcal ME/kg, as determined in accordance with Regulation PF9.

Steps to calculate nutrient requirement and to formulate a balanced diet for dogs and cats at different stages.

Step 1

Estimate the weight of the pet

Step 2

Determine the nutritional needs of the pet

Steps 3

Ascertain the nutritive value of common feed.

Step 4

Determine how much feed each dog or cat can consume daily. This enables us to calculate if certain feeds can be consumed in sufficient quantities to provide the require nutrition.

Step 5

Calculate the feed required for each dog or cat.

Metabolizable energy requirements for growth of puppies after weaning[a,b]

$$ME\ (kcal) = maintenance\ amount \times 3.2 \times [e^{(-0.87p)} - 0.1]$$

$$ME\ (kcal) = 130 \times BW_a^{0.75} \times 3.2 \times [e^{(-0.87p)} - 0.1]$$

Where: $p = BW_a / BW_m$; BW_a = actual body weight at time of evaluation (kg); BW_m = expected mature body weight (kg); e = base of natural log = 2.718

[a] This table refers to puppies after weaning.

[b] Maintenance energy requirement of inactive puppies (such as pet puppies without opportunity and / or stimulus to exercise) may be lower by 10 to 20 percent, and maintenance energy requirements of very active puppies, such as Great Danes in Kennels, may be higher; Source: Nutrient requirements of dogs and cats (NRC, 2006)

Example:

Labrador puppy 16 weeks of age, 17 kg actual body weight, expected mature body weight 35 kg.

$$ME\ (kcal) = 130 \times BW_a^{0.75} \times 3.2 \times [e^{(-0.87p)} - 0.1]$$

$$ME\ (kcal) = 130 \times 17^{0.75} \times 3.2 \times [e^{(-0.87 \times 17/35)} - 0.1] = 1934\ kcal$$

Table 7. Average energy requirements during growth in dogs (FEDIAF, 2011)

Puppies*	Age	Energy requirement	
Rapidly growing	Newborn puppies upto 50% of adult weight	25 kcal/100 g BW 210 kcal/$kg^{0.75}$	105 kJ/100 g BW 880 kJ/$kg^{0.75}$
Actively growing	50 to 80% of adult weight	175 kcal/$kg^{0.75}$	730 kJ/$kg^{0.75}$
Growing	80 to 100% of adult weight	140 kcal/$kg^{0.75}$	kJ/$kg^{0.75}$

***Overfeeding in puppies can result in skeletal deformities especially in large and giant breeds. Therefore, puppies should never be fed ad libitum and weight gain be closely monitored.**

Table 8. Daily metabolizable energy requirements for adult dogs at maintenance*

Type	Kcal x /$kg^{0.75}$
Average for laboratory kennel dogs or active pet dogs[a]	130
Above average requirements:	
Young adult laboratory dogs or young adult active pet dogs	140
Adult laboratory Great Danes or active pet Great Danes	200
Adult laboratory terriers or active pet terriers	180
Below average requirements:	
Inactive pet dogs[b]	95
Older laboratory dogs or older active pet dogs or laboratory Newfoundlands	105

[a] Dogs kept in a domestic environment with strong stimulus and ample opportunity to exercise, such as dogs in multiple dog house-holds in the country or in a house with a large yard;

[b] Dogs kept in a domestic environment with little stimulus and opportunity to exercise. Requirements of older or overweight dogs may still be overestimated;

* Nutrient requirements of dogs and cats (NRC, 2006)

Table 9. Practical recommendations for Maintenance energy requirements (MER) in dogs at different ages [Source: FEDIAF (2011)]

Age (Years)	Average	
	kcal ME/$kg^{0.75}$	kJ ME/$kg^{0.75}$
1-2	130	550
3-7	115	480
> 7 (senior dogs)	100	418
Obese prone adults	90	376
Breed specific differences:		
Great Danes	200	837
New foundlands	105	439

Table 10. Recommendations for daily energy requirements (DER) of adult dogs in relation to activity

Activity level	kcal ME/$kg^{0.75}$	kJ ME/$kg^{0.75}$
Low activity (< 1h/day) (e.g. walking)	100	418
Moderate activity (1-3 h/day) (e.g. playing)	125	523
High activity (3-6 h/day) (working dogs, e.g. sheep dogs)	150 - 175	628 - 732
High activity under under extreme conditions (racing sled dogs 168 km/d in extreme cold)	860 - 1240	3600 - 5190

Table 11. Recommended Allowance (RA; g per kg metabolic body size) of growing puppies, adults and bitches in late gestation and peak lactation*

Nutrient	Growing puppies, Age 4-14 weeks	Growing puppies, Age 14 week and older	Adults	Late gestation & peak lactation
Crude protein, g	15.7	12.2	3.28	24.6*
Total fat, g	-	5.9	1.8	10.5
Linoleic acid, g	-	0.8	0.36	1.6
-Linolenic acid, g	-	0.05	0.014	0.10
Arachidonic acid, g	-	0.022	-	-
EPA &DHA,g	-	0.036	0.03	0.06
Calcium, g	-	0.68	0.13	0.82
Phosphorus, g	-	0.68	0.10	0.58

*** Adapted from Nutrient requirements of dogs and cats (NRC, 2006); [a] For small bitches suckling two puppies: 10 g/kg$W^{0.75}$; for medium sized bitches suckling six puppies: 20 g/kg$W^{0.75}$; for large bitches suckling eight puppies: 20 g/kg$W^{0.75}$**

Daily metabolizable energy requirements for bitches in late gestation (4 weeks after mating until parturition)*

ME (kcal) = maintenance + 26 kcal x kg BW

Average maintenance requirements 130 kcal x kg $BW^{0.75}$

ME (kcal) = 130 kcal x kg $BW^{0.75}$ + 26 kcal x kg BW

***Nutrient requirements of dogs and cats (NRC, 2006)**

Example:

Body weight of bitch 22 kg

Maintenance requirements = 130 kcal x $22^{0.75}$ = 130 x 10.16 = 1,320 kcal

Requirements for gestation 22 x 26 kcal = 572 kcal

Total requirements 1,320 kcal + 572 kcal = 1,892 kcal

Daily metabolizable energy requirements for lactating bitches based on number of puppies and weeks of lactation

Requirements for lactation

ME (kcal) = maintenance + BW x (24n + 12m) x L

Extrapolated maintenance energy requirements during lactation: 145 kcal x $BW^{0.75}$

Average maintenance requirements 130 kcal x kg $BW^{0.75}$

ME (kcal) = 145 kcal x $BW^{0.75}$ + BW x (24n + 12m) x L

Where:

BW = body weight of bitch (kg)

n = number of puppies between 1 and 4

m = number of puppies between 5 and 8 (<5 puppies m = 0)

L = correction factor for stage of lactation: week 1, 0.75; week 2, 0.95; week 3, 1.1; and week 4,1.2*,**

* Expressed as a percentage of the mean daily milk yield during the first 4 weeks of lactation, milk production is as follows: 75 percent in week 1, 95 percent in week 2, 110 percent in week 3 and 120 percent in week 4.

** From the nutrient content of bitch milk, the gross energy content is estimated to amount to about 1.45 kcal per g wet weight. The energy content of canine milk shows little variation except for the colostral period. Milk yield has been estimated to amount up to 8 percent of body weight. These studies agree that peak lactation in the bitch occurs in the 4th week; Nutrient requirements of dogs and cats (NRC, 2006)

Example:

Bitch 22kg, 6 puppies third week of lactation

Maintenance requirements = 145 kcal x $22^{0.75}$ = 145 x 10.16 = 1,473 kcal

Number of puppies = 6; n=4, m=2; Stage of lactation-third week: L = 1.1

Requirements for lactation = 22 x (24 x 4 + 12 x 2) x 1.1 kcal = 2,904 kcal

Total requirements = 1,473 kcal + 2,904 kcal = 4,377 kcal

Table 12. Average energy requirements during reproduction in dogs*

Bitches	Age	Energy requirement	
Gestation	First 4 weeks of gestation	132 kcal/kg$^{0.75}$	550 kJ/kg$^{0.75}$
	Last 5 weeks of gestation	132 kcal/kg$^{0.75}$ + 26 kcal/kg BW	550 kJ/kg$^{0.75}$+ 110 kJ/ kg BW
Lactation	1 to 4 puppies	132 kcal/kg$^{0.75}$ + 24n x kg BW x L	550 kJ/kg$^{0.75}$ + 100n x kg BW x L
	Lactating bitch, 5 to 8 puppies	132 kcal/kg$^{0.75}$ + (96 + 12n) x kg BW x L	550 kJ/kg$^{0.75}$ + (400 + 50n)x kg BW x L

***Source: FEDIAF (2011); n = number of puppies; L = 0.75 in week 1, 0.95 in week 2, 1.1 in week 3 and 1.2 in week 4.**

Daily metabolizable energy requirements for growth of kittens after weaning*

ME (kcal) = maintenance amount x 6.7 x $[e^{(-0.189p)}-0.66]$

ME (kcal) = 100 x $BW_a^{0.67}$ x 6.7 x $[e^{(-0.189p)}-0.66]$

Where:

p= BW_a/ BW_m; BW_a= actual body weight at time of evaluation (kg); BW_m= expected mature body weight (kg); e = base of natural log = 2.718

* Nutrient requirements of dogsandcats (NRC, 2006)

Example:

Kitten, 1 kg BW_a, 4 kg BW_m

ME (kcal) = 100 x $1^{0.67}$ x 6.7 x $[e^{(-0.189 \times 1/4)} - 0.66]$ = 198 kcal

Table 13. Average energy requirements during growth in cats (FEDIAF, 2011)

Kittens	Age	Times MER
Rapidly growing	Upto 4 months	2.0-2.5
Actively growing	4 to 9 months	1.75-2.0
Growing	9 to 12 months	1.5

Table 14. Recommended Allowance (RA; g per kg metabolic body size) of growing kittens, adults and queens in late gestation and peak lactation*

Nutrient	Growing kittens	Adults	Late gestation	Peak lactation
Crude protein, g	11.8	4.96	7.40	16.10
Taurine, g	0.021	0.0099	0.018	0.018
Total fat, g	4.7	2.2	-	4.8
Linoleic acid, g	0.29	0.014	-	0.3
-Linolenic acid, g	0.010	-	-	0.011
Arachidonic acid, g	0.001	0.0015	-	0.011
EPA &DHA,g	0.005	0.0025	-	0.0044

Nutrient	Growing kittens	Adults	Late gestation	Peak lactation
Calcium, g	0.410	0.071	-	0.565
Phosphorus, g	0.372	0.063	-	0.411

*** Adapted from Nutrient requirements of dogs and cats (NRC, 2006)**

Table 15. Average daily energy requirements of adult cats (FEDIAF, 2011)

Gender / Age	kcal ME/kg$^{0.67}$	kcal ME/kg BW	kJ ME/kg$^{0.67}$	kJ ME/kg BW
Intact male & female	80	50-60	335	210-250
Neutered and indoor cats	52-87	35-55	215-365	145-230
Active cats	100	60-70	418	250-290

Table 16. Average energy requirements during reproduction in cats (FEDIAF, 2011)

Queens	Reproduction phase		
Gestation		145 kcal/kg$^{0.67}$	585 kJ/kg$^{0.67}$
Lactation	< 3 kittens	100 kcal/kg$^{0.67}$ + 18 x kg BW x L	418 kJ/kg$^{0.67}$+ 75 x kg BW x L
	3-4 kittens	100 kcal/kg$^{0.67}$ + 60 x kg BW x L	418 kJ/kg$^{0.67}$+ 250 x kg BW x L
	> 4 kittens	100 kcal/kg$^{0.67}$ + 70 x kg BW x L	418 kJ/kg$^{0.67}$+ 293 x kg BW x L

***Source: FEDIAF (2011); L = 0.9 in weeks 1-2 of lactation; 1.2 in weeks 3-4; 1.1 in week 5; 1 in week 6 and 0.8 in week 7.**

Table 17. Daily metabolizable energy requirements for lactating queens*

Kittens	Energy requirement
< 3 kittens	ME kcal = maintenance+ 18 x kg BW x L ME kcal = 100 kcal/kg$^{0.67}$ + 18 x kg BW x L
3-4 kittens	ME kcal = maintenance+ 60 x kg BW x L ME kcal = 100 kcal/kg$^{0.67}$ + 60 x kg BW x L
> 4 kittens	ME kcal = maintenance+ 70 x kg BW x L ME kcal = 100 kcal/kg$^{0.67}$ + 70 x kg BW x L

***Nutrient requirements of dogs and cats (NRC, 2006); L = factor for stage of lactation from week 1 to week 7: 0.9, 0.9, 1.2, 1.2, 1.1, 1 and 0.8**

Example:

Queen, 3.5 kg BW at mating, 4 kittens, peak lactation (week 3)

ME kcal = 100 x 3.5$^{0.67}$ + (60 x 3.5 x 1.2) = 231 + 252 = 483 kcal

Table 18. Composition of ingredients (unit per 100 g) used for dogs and cats

Ingredients	Energy (kcal)	Protein (g)
Milk	67	3.2
Egg	143	11.3
Wheat	346	11.8
Pulses	350	22.5
Beans	49	4.7
Meat	194	19
Rice	345	6.8
Carrot	48	0.9

Table 19. Starch content of some common plant ingredients used in dog and cat diets

Product	Starch content (g/kg diet DM)
Barley	500-516
Oats	400
Sorghum	620-741
Maize	630-728
Rye	520
Wheat	580-621
Peas	410
Horse beans	350
Pearl barley product (2% crude fibre)	530
Pearl barley product (7.5% crude fibre)	350
Pearl barley product (10.5% crude fibre)	300
Hominy feed	370
Maize bran	260
Rice bran (< 3% husks)	270
Rice bran (3-10 % husks)	230
Rice	810
Wheat feed flour	500
Wheat middlings	220
Wheat bran	140
Wheat germ meal	280
Dried potatoes	650
Dried potato pulp	280
Corn gluten feed	190

Table 20. Safe upper limits of selected carbohydrates for adult dog and cat maintenance diets (g/kg diet, DM basis)*

Item	Dogs	Cats
Glucose	ND	50-150
Sucrose	350	50-150
Lactose	100	50
Cooked corn starch	ND	240
Cooked corn flour	436	ND
Cooked potato flour	504	ND
Cooked rice flour	441	ND
Cooked wheat flour	491	ND
Raffinose / stachyose	50	50
Fructooligosaccharides	40	7.5
Inulin	70	ND
Mannanoligosaccharides	5.9	ND
Xylooligosaccharides	5	ND
Transgalactooligosaccharides	5.9	ND
Cellulose	94	100
Guar gum	34	ND
Pectin	34	ND
Bet pulp	75	ND
Wheat bran	128	100
Peanut hulls	67	ND
Tomato pomace	87	ND
Oat fibre	75	ND
Fibre blend	83	83

Table 21. Estimated daily food allowance for Cats

Cat	Weight of Cat (kg)	Food intake (90% DM; ME 3.2 kcal/g)	
		g/kg BW	g/Cat
Kitten			
10 weeks	0.9-1.1	78	70-86
20 weeks	1.9-2.5	41	78-103
30 weeks	2.5-3.8	31	78-118
40 weeks	2.9-3.8	25	73-95

Cat	Weight of Cat (kg)	Food intake (90% DM; ME 3.2 kcal/g)	
		g/kg BW	g/Cat
Adult (50 weeks of age or older)**			
Inactive	2.2-4.5	22	48-90
Active	2.2-4.5	25	55-113
Gestation	2.5-4.0	31	78-124
Lactation*	2.5-4.0	25	172-312

* Queens nursing 4-5 kittens in week 6 of lactation;

** During pregnancy 1.25 to 1.5 times of maintenance ration can be given till the end of gestation; during lactation, it should be increased to 2 to 3.5 times of adult maintenance ration.

Table 22. Nutrient content of dry, semi-moist and canned dog foods

	As fed basis	DM basis
Dry		
Moisture (%)	6-10	0
Fat (%)	7-20	8-22
Protein (%)	16-30	18-32
Carbohydrate (%)	41-70	46-74
ME (kcal/kg)	2800-4050	3000-4500
Semi-moist		
Moisture (%)	15-30	0
Fat (%)	7-10	8-14
Protein (%)	17-20	20-28
Carbohydrate (%)	40-60	58-72
ME (kcal/kg)	2550-2880	3000-4000
Canned		
Moisture (%)	75	0
Fat (%)	5-8	20-32
Protein (%)	7-13	28-50
Carbohydrate (%)	4-13	18-57
ME (kcal/kg)	875-1250	3500-5000

Formulation of balance ration for dogs and cats

1. Growing puppy

Problem 1. Calculate the nutrient requirement of 5 kg growing (four months of age) German Shephard (Alsatian) dog and formulate a balanced diet.

Solution:

Metabolic body weight: $5^{0.75} = 3.54$ kg

Energy requirement (up to 50% of adult weight) = 210 kcal ME/kg$^{0.75}$

Protein requirement (14 weeks and older puppies) = 12.2 g/kg$^{0.75}$

Requirements per day: Energy: 3.54 x 210 = 702 kcal ME

Protein: 3.54 x 12.20 = 40.8 g CP

Composition of ingredients, per 100 g			Balanced diet I*			Balanced diet II*		
Energy, kcal	Protein, g	Ingredient, g	Quantity, g	Energy, kcal	Protein, g	Quantity, g	Energy, kcal	Protein, g
67	3.2	Milk	300	201	9.60	200	134	6.40
143	11.3	Egg	100	143	11.3	50	71.5	5.65
346	11.8	Wheat	50	173	5.90	-	-	-
350	22.5	Pulses	50	175	11.25	-	-	-
49	4.7	Beans	50	24.5	2.35	-	-	-
194	19	Meat	-	-	-	125	242.5	23.75
345	6.8	Rice	-	-	-	75	258.75	5.10
				716.5	**40.4**		**706.75**	**40.90**

***Minerals and vitamins supplement need to be provided daily to meet their requirements.**

Problem 2. Calculate the nutrient requirement of 2 kg growing (four months of age) Spitz dog and formulate a balanced diet.

Solution:

Metabolic body weight: $2^{0.75} = 1.68$ kg

Energy requirement (up to 50% of adult weight) = 210 kcal ME/kg$^{0.75}$

Protein requirement (14 weeks and older puppies) = 12.2 g/kg$^{0.75}$

Requirements per day: Energy: 1.68 x 210 = 353 kcal ME

Protein: 1.68 x 12.20 = 20.52 g CP

Composition of ingredients, per 100 g			Balanced diet I*			Balanced diet II*		
Energy, kcal	Protein, g	Ingredient, g	Quantity, g	Energy, kcal	Protein, g	Quantity, g	Energy, kcal	Protein, g
67	3.2	Milk	120	80.4	3.84	200	134	6.40
143	11.3	Egg	50	71.5	5.65	20	35.75	2.83
346	11.8	Wheat	40	138.4	4.72	-	-	-
350	22.5	Pulses	20	70.0	4.50	-	-	-

Composition of ingredients, per 100 g			Balanced diet I*			Balanced diet II*		
Energy, kcal	**Protein, g**	**Ingredient, g**	**Quantity, g**	**Energy, kcal**	**Protein, g**	**Quantity, g**	**Energy, kcal**	**Protein, g**
49	4.7	Beans	40	19.6	1.88	-	-	-
194	19	Meat	-	-	-	50	97.0	9.50
345	6.8	Rice	-	-	-	25	86.25	1.70
				379.9	**20.59**		**353.00**	**20.43**

***Minerals and vitamins supplement need to be provided daily to meet their requirements.**

2. Adult dogs

Problem 3. Calculate the nutrient requirement of 25 kg adult (four years of age) German Shephard (Alsatian) dog and formulate a balanced diet.

Solution:

Metabolic body weight: $25^{0.75} = 11.18$ kg

Energy requirement (3-7 years of age) = 115 kcal ME/kg$^{0.75}$

Protein requirement (adults) = 3.28 g/kg$^{0.75}$

Requirements per day: Energy: 11.18 x 115 = 1286 kcal ME

Protein: 11.18 x 3.28 = 37.0 g CP

Composition of ingredients, per 100 g			Balanced diet I*			Balanced diet II*		
Energy, kcal	Protein, g	Ingredient, g	Quantity, g	Energy, kcal	Protein, g	Quantity, g	Energy, kcal	Protein, g
67	3.2	Milk	275	184.25	8.80	150	100.5	4.80
345	6.8	Rice	280	966	19.04	250	862.5	17.0
350	22.5	Pulses	25	87.5	5.63	-	-	-
48	0.9	Carrot	50	24.0	0.45	-	-	-
49	4.7	Beans	60	29.4	2.82	-	-	-
346	11.8	Wheat	-	-	-	70	242.2	8.26
194	19	Meat	-	-	-	40	77.6	7.60
				1291.5	**36.74**		**1282.8**	**37.66**

Problem 4. Calculate the nutrient requirement of 8 kg adult (four years of age) Spitz dog and formulate a balanced diet.

Solution:

Metabolic body weight: $8^{0.75} = 4.76$ kg

Energy requirement (3-7 years of age) = 115 kcal ME/kg$^{0.75}$

Protein requirement (adults) = 3.28 g/kg$^{0.75}$

Requirements per day: Energy: 4.76 x 115 = 547 kcal ME

Protein: 4.76 x 3.28 = 15.6 g CP

Composition of ingredients, per 100 g			**Balanced diet I***			**Balanced diet II***		
Energy, kcal	**Protein, g**	**Ingredient, g**	**Quantity, g**	**Energy, kcal**	**Protein, g**	**Quantity, g**	**Energy, kcal**	**Protein, g**
67	3.2	Milk	100	67.0	3.20	100	67.0	3.20
350	22.5	Pulses	20	70.0	4.50	-	-	-
48	0.9	Carrot	20	9.6	0.18	-	-	-
194	19	Meat	-	-	-	25	48.5	4.75
345	6.8	Rice	120	414	8.16	125	431.25	8.50
				560.6	16.04		546.75	16.45
194	19	Meat	-	-	-	40	77.6	7.60
				1291.5	**36.74**		**1282.8**	**37.66**

***Minerals and vitamins supplement need to be provided daily to meet their requirements.**

A balanced ration for Border Security Force dogs

Ingredient	Quantity per day*
Rice / Wheat grain	480 g
Dog biscuit	30 g
Vegetables, fresh	230 g
Meat dressed with bones	680 g

***Dogs are supplemented with calcium gluconate tablet, multivitamin tablet and shark liver oil as per the need.**

Feeding schedule for dogs: A guide chart

Body weight (kg)	Cereals (g)	Meat* (g)	Milk (g)
2-5	100-200	50-100	125-150
5-10	200-300	100-200	150-250
10-20	300-500	200-400	250-400

***The 'meat' portion of the diet may be completely avoided by including 1-2 eggs and pulses / oilseed meals. Vegetables may also be added to meet the requirement of fibre and micronutrients.**

A standard balanced dog food should contain 20 to 22 % CP, 4 to 5 % fat, 5 to 6 % fibre and 4 to 5 % ash on dry matter basis. ME content ranges from 3.2 to 4.0 Mcal per kg food on dry matter basis.

Feeding Schedule for Cats : A Guide Chart

Cat	Weight of Cat (kg)	Food intake (DM 90%, CP 25.2 %, Fat 8.1 %, ME 3.2 Kcal/g)	
		g/kg BW	g/Cat
Kitten			
10 Weeks	0.9-1.1	78	70-86
20 Weeks	1.9-2.5	41	78-103
30 Weeks	2.5-3.8	31	78118
40 Weeks	2.9-3.8	25	73-95
Adult (50 Weeks of age or older)			
Inactive	2.2-4.5	22	48-9
Active	2.2-4.5	25	55-113
Gestation	2.5-4.0	31	78-124
Lactation*	2.5-4.0	78	172-312

***Queens nursing 4-5 kittens in week 6 of lactation.**

Growing kitten

Problem 1. Calculate the nutrient requirement of 1 kg growing active cat (six months of age) and formulate a balanced diet.

Solution:

Metabolic body weight: $1^{0.67} = 1$ kg

Energy requirement: Maintenance energy requirement (MER) for active cat = 100 kcal ME/kg$^{0.67}$

ME requirement for cat of 4-9 months = 1.75 to 2.0 times of MER

Protein requirement (growing kittens) = 11.80 g/kg$^{0.67}$

Requirements per day: Energy: 2.0 x 100 = 200 kcal ME

Protein: 1.0 x 11.80 = 11.80 g CP

Composition of ingredients, per 100 g			Balanced kitten diet*		
Energy, kcal	**Protein, g**	**Ingredient, g**	**Quantity, g**	**Energy, kcal**	**Protein, g**
67	3.2	Milk	130	87.10	4.16
194	19	Meat	35	67.90	6.65
345	6.8	Rice	15	51.75	1.02
		Total		**206.75**	**11.83**

***Minerals and vitamins supplement need to be provided daily to meet their requirements.**

2. Adult cats

Problem 2. Calculate the nutrient requirement of 3 kg active adult cat (three years of age) and formulate a balanced diet.

Solution:

Metabolic body weight: $3^{0.67} – 2.09$ kg

Energy requirement (active cat) = 100 kcal ME/kg$^{0.67}$

Protein requirement (adults) = 4.96 g/kg$^{0.67}$

Requirements per day: Energy: 2.09 x 100 = 209 kcal ME

Protein: 2.09 x 4.96 = 10.37 g CP

Composition of ingredients, per 100 g			Balanced cat diet*		
Energy, kcal	**Protein, g**	**Ingredient, g**	**Quantity, g**	**Energy, kcal**	**Protein, g**
67	3.2	Milk	140	93.80	4.48
194	19	Meat	25	48.50	4.75
345	6.8	Rice	20	69.00	1.36
Total				**211.30**	**10.59**

*** Minerals and vitamins supplement need to be provided daily to meet their requirements.**

CHAPTER 40

FEEDS AND FEEDING SCHEDULE OF ZOO ANIMALS AND BIRDS-DIET CHARTS

The science of nutrition is an integral component to the management of zoo animals and must be addressed in a scientific way. The challenge of a zoo nutrition programme is to provide a nutritionally balanced diet, which reasonably stimulates natural feeding behaviors. The goal of zoo nutrition programme differs vastly from those of livestock or pet nutrition. Its objective includes the optimum reproduction, longevity and behavioral normality by an attempt to meet physiological and psychological needs of the species that have evolved under diverse environmental circumstances, to occupy specialized niches with ability to make diet choices according to season, physiology, environment or individual requirements.

Nutrients requirement of zoo animals

The most obvious challenge in providing optimum nutrition to captive wildlife is the lacking of known requirement of nutrients. In most of the zoos animals are fed empirically based upon whether they are herbivore, omnivore or carnivore. A useful approach when formulating diets for wild animals is to consider dietary habits in the wild, oral and gastrointestinal morphology and physiological needs of similar species whose requirements are known, and environmental features that affect energy and nutrient need. Most of the information regarding nutrition of zoo animals available has been emanated from research conducted abroad. There are three main sources of information, which are used to develop zoo animal diets:

1. The documented/known natural food habit of the animal, which often does not contain specific ingredients.
2. Record from other captive management facilities, which are often incomplete or developed by trial and error methods.

3. Data from related domesticated animal models.

The nutrient requirement of any individual depends on its species, age sex, body size, physiological state, activity, health status and other environmental factors. Animal nutritionist usually express nutrient requirement as percentage or as concentration, however, it seems more precise to express it on the basis of energy density, since under normal circumstances animal will adjust its feed intake to meet their calorie requirements. It very difficult to achieve precision when feeding zoo animals because of uncertainty of level of feed intake. Under normal circumstances, animal will not consume energy in excess of their requirement. However, obesity is quite common in captive zoo animals due to limited activity. Zoo animals (especially herbivore) in a group or herd tend to fed in excess of their requirement.

Points to be considered while preparing zoo diets

Special attention should be paid to the major nutrients like energy and protein content of diet. Always ensure availability of important minerals and vitamins for the animals in captivity. Following points should be kept in consideration while preparing diets for herbivores, carnivores and omnivores zoo animals.

A. Herbivores

1. Good quality roughage (greens or hay) should be included in the diet of herbivores for optimum microbial fermentation.
2. Concentrate/pellet must be formulated by an animal nutritionist with the peculiarities of species and physiological status of animal. Amount of concentrate should not exceed 40% of daily feed intake.
3. Large herbivores like elephant (post-gastric fermenter) consume large quantities of fibrous feed due to hind-gut fermentation and rapid transit time, which results in poor digestibility.
4. Feed ingredients to be used for formulation of concentrate must be detoxified for their toxic principles (protease inhibitors in soybean, gossypol in cottonseed, aflatoxins in grains, etc.).
5. When using top feed/browse for behavioral stimulation, care must be taken for anti-nutritional/toxic factors present in fodder (cyanide in sorghum and Jowar, oxalates in hybrid Napier, guinea and Para grass, nitrates or nitrites in oats, saponins in legume fodders, etc.).
6. Seasonal available fodders and tree leaves may be gradually incorporated in diet.

B. Carnivores

1. Carnivores require highly digestible and high quality food resources with well balance composition of all the types of nutrients.

2. Carnivores are major attraction of zoo and need unique type of nutrition. Felids need a dietary requirement for taurine amino acid, preformed vitamin A (retinol) and arachidonic acid fatty acid.
3. They poorly utilize cellulose and unboiled starch. Thus their diet should not contain more than 3-4% CF.
4. Felid derive their most of energy and blood glucose from catabolism of amino acid rather than soluble carbohydrates through gluconeogenesis. Thus, their diet must contain higher proportion of protein (minimum 21-22 %) with limited amount of carbohydrates. Their kittens require a minimum of 28% protein in diet for growth purposes.
5. To stimulate intermittent feeding behavior weekly off day may be practiced.
6. While offering fishes, thiamine and vitamin E should be supplemented as fishes contain thiaminase enzyme which destroys thiamine.
7. Commercial canned, frozen or dried diets formulated for dog or cat may also be provided. Prey items like rat, mice, chick, fish or invertebrate insects should also be supplied on regular basis for their enrichment.
8. As meat is deficient in calcium therefore Ca: P ratio is highly imbalanced. Due to old age and easy access to food in captivity many felid do not chew bones. Metabolic bone disorders are quite common in large cats. Hence, supply 100 mg Ca/kg body weight/day.
9. Young felids can be hand reared by providing powdered commercial milk replacers with boiled water in the ratio of 1:1or 1:2 in the first week to 9th week.

C. Omnivores

1. This is the largest group comprising of birds, apes, pigs, rodents, bats, marsupials, sloths and bears. Their gastro intestinal tract falls in between that of herbivores and carnivores.
2. Birds require a larger quantity of feed in proportion to their body weight. They feed on seeds, fruits, nectar, sap, insects, reptiles, etc. They generally eat their food to satisfy their gut volume and energy requirements.
3. Birds require high energy (fat) food during cold winter to maintain their high body temperature. Proteins may be utilized for energy purpose (gluconeogenesis) in carnivore and insectivore birds.
4. All birds require essential fatty acid. However, sufficient linoleic acid (1%) meets the requirement of all EFA.
5. In avian, minerals like Cu and Fe play crucial role in maintaining coat colour. Shell grits and enzymes (xylanase, phytase and glucanase) must be supplied in the diet.

6. Some of the marsupials are herbivores (kangaroo) while others are (Tasmanian devil) primarily subsist on carnivore diet. Captive marsupials are maintained on a wide variety of diets which include grain meal, hay, green plants, bred, fruit, chopped meat, mealworm, insects, spiders and birds.
7. Chiropterans or bats are voracious feeders. A small fruit bat may consume fruit more than its body weight at a single meal; insectivorous bat usually capture their prey while in flight and may be fed on termites, cockroaches, spiders, woodlice, beetles, moths, grasshopper, mosquito etc. However, Vampire bats feed on blood of cattle, sheep or chicken. Blood should be defibrinated by stirring and is stored at 4°C until it is given to the bats in the bowls late in the afternoon. Live chickens may be also be placed in enclosure so that bats may directly suck the blood.
8. Zoo reptiles mostly feed on carnivore's diet which includes fish rat, mice, birds, small rabbits and amphibians. Any change in environmental temperature leads to change in their feeding behaviour. Higher temperature increases digestion and metabolic rates. Other factors such as photoperiod, humidity, type of food population density also play vital role on the feeding behaviour of reptiles.

Feeding of zoo animals in different conditions

A. Feeding of sick animal

1. Most seriously ill-wild animals either refuse to eat or just pick at the food. Force feeding often may not be recommended especially in carnivores or herbivores.
2. Provide clean drinking water *ad-libitum* to all the sick wild animals in general.
3. Soft diet like chicken preferably in cooked form (soup) may be attempted as one of the significant diets in wild animal species.
4. Intra-venous fluids like dextrose saline, ringers lactate etc may be chosen in wild animals that have totally ceased eating activities.
5. Attempts should be made to provide the most palatable food for the targeted wild species.
6. Carnivores with severe gastritis tend to vomit often. In such cases, a blend diet should be tried. In gastritis affected wild animals, in addition to administration of 5% dextrose saline solution, egg white may be offered.
7. Provide drinking water *ad-libitum* to all the sick wild animals in general.
8. Provision of bland diet in small quantities spaced in several daily doses may helpful to minimize the incidences of vomiting or rejection by the concerned wild animal species.

B. Feeding of geriatric animal

1. Avoid feeding with less digestible feeds.
2. Avoid full stomach feeding.
3. Provide adequate quantities of vitamins and minerals.
4. Provide *ad-libitum* water.

Table No. 1. Generalized feeding strategy for different classes of zoo animals

Nutrients	Requirement			
	Carnivore	Omnivore	Ruminant	Hindgut fermenter
Carbohydrate	Minimal or nil Very low fibre	> 50% of diet low fibre	High roughage Low starch	30-40% starch 8-10 % of roughage
Protein	High (20-30%)	12-18 %	Low	8-10%
Fat	High (20-30%)	4-5%	Low	Low (>2%)
Energy source	Protein and fat	Carbohydrate and fat	Fibre (VFAs)	Carbohydrate
Vitamin	Highly required (all)	moderately	A, D_3 and E	A, D_3 and E
Minerals	Highly required (all; macro minerals)	moderately	Low	Low

Feeding schedule of zoo animals

The feeding schedules of zoo animals vary widely which depends on species, age, sex and physiological status (young, adult, old, pregnant and lactating) of animal under consideration. Feeding schedule also differs from place to place due to availability and nutritive composition of different feed ingredients. A brief idea regarding the feed and feeding schedule of common zoo animals have been provided below.

1. Feeding schedule for common zoo herbivores

Species	Mash (kg)	Green fodder (kg)	Tree fodder (kg)
Deer			
Sambar	2.0	5.0	-
Swamp deer	2.0	3.0	-
Sangai	1.0	2.0	5.0
Chital	0.75	2.0	-
Hog deer	0.50	3.0	2.0

Species	Mash (kg)	Green fodder (kg)	Tree fodder (kg)
Antelopes and wild goats			
Gaur	3.0-4.0	*Ad lib.*	5.0-6.0
Nil gai	2.0-2.5	*Ad lib.*	3.0-4.0
Chinkara	0.5-1.0	2.0-3.0	2.0-3.0
Black buck	0.5-1.0	2.0-3.0	2.5-3.5
Goral	1.5-2.0	5.0	1.0

Note: In elephant, 400 kg green fodder along with 6 kg roti, 50 g oil, 250 g salt and 500 g jaggery. In Rhinoceros, 200 kg green fodder along with 2 kg g, 1kg wheat bran, 1 kg beans, 10 no. bananas and 40 g salt.

2. Feeding schedule for common zoo carnivores

Species	Meat	Milk	Egg (no.)	Chicken	Remarks
Tiger	8.0-10.0	0.5	1.0	weekly	Monday weekly off
Lion	8.0-10.0	-	-	-	-do-
Leopard	4.0	-	-	0.75	-do-, 250 g liver
Jungle cat	0.5	-	-	-	-do-
Leopard cat	0.5	-	-	-	-do-
Fishing cat	0.5	0.2	1.0	0.2	-do-
Golden cat	1.0	-	-	-	-do-
Wild dog	2.5	0.25	-	weekly	-do-,250 g liver
Jackal	2.0	-	-	-	-do-
Wolf	4.0	-	-	-	-do-
Fox	0.5	-	-	-	-do-

Note: In bears have diverse feeding habit. They mostly do well on canine diets and fish. The feeding regimen should be diverse as much as possible. Their feed may consist of bread 250 g, banana 2 kg, egg 1-2 no., vegetables 2 kg, rice 500 g, milk 500 ml jaggery or honey 200 g, mineral mixture should be supplemented on meat based diets. In Red panda, milk 1 l, sugar 20 g, carrot 200 g, variety of fruit and vegetables 2 kg, egg 2 nos., bread 150 g and bamboo leaves 2 kg.

3. Feeding schedule for zoo birds

Majority of birds are seed eaters. The varieties of seeds include agricultural cereals, grass seeds wild tree seeds and oil seeds. Many like Koel, sparrow, quails peacock, falcon are insectivores and rely on reptiles also. All birds should be supplemented with vitamin and mineral mixture 2 g/day.

4. Feeding schedule for zoo reptiles

Their natural diets consist of rodents, birds, amphibians, fish and eggs.

i. Crocodiles and Gharials

In zoo they should be offered with intact preys like whole fish or mice. The viscera along with strips of meat may also be given. They feed better when dietary ingredients are tossed to them just lateral to the mouths.

ii. Snakes

King cobra is almost strict snake eater. Boas and python prefer to feed on warm blooded preys (rodents and birds). Snakes should be fed once or twice in fortnight/ month. As general rule large reptiles should be fed less frequently e.g. python should be fed once every four to six weeks.

Feeding of wild animals

Herbivores	
Barking deer/Spotted deer/Sambar • Wheat bran 3 kg • White Bengal gram 25 gm • Grass 7 kg • Green leaves 3 kg • Salt 10 gm • Stylo/Lucerne 2 kg	**Zebra (adult)** • Wheat bran 2 kg • White Bengal gram 100 gm • Grass and leaves 30 kg • Mineral mixture 10 days in a month at regular interval • Lucerne 5 kg
Indian Elephant • Ragi 7 kg • Horse gram 2 kg • Salt 200 gm • Jaggery 250 gm • Grass 200 kg • Sugarcane 4 nos • Green tree leaves– Bamboo/Stylo 250 kg • Coconut 500 gm (1 no.)	**Camel** • Wheat bran 2 kg • White Bengal gram 500 gm • Grass 30 kg • Leaves (neem, bamboo, subabul, bauhina) 20 kg • Salt 50 gm • Cabbage 2 kg • Banana 6 no. • Leaves of acacia 5 kg
Indian Bison (Gaur) • Wheat bran 3 kg • Bengal gram 0.5 kg • Green leaves / grass L.S.	

Carnivores	
Lion and Tiger • Beef (with bone) 10 kg • Liver 500 gm • Chicken 1 kg (weekly once)	**Panther** • Beef (with bone) 3 kg • Liver 250 gm • Chicken 1 kg (weekly once)
Porcupine • Rice 100 gm • Carrot 100 gm • Cabbage 100 gm • Tapioca or sweet potato 100 gm • Soya bean 20 gm • Groundnut (WOS) 150 gm	**Sloth bear** • Ragi (cooked) 250 gm • Rice gruel + black gram 150+50 gm • Orange / Mango 2 no • Guava 4 no • Tapioca 100 gm • Jaggery 200 gm • Groundnut (WOS) 100 gm
Jackal • Beef (without bone) 500 gm • Liver 100 gm • Chicken 250 gm (weekly once)	**Hyena** • Beef 3 kg/day • Liver 100 gm • Chicken 500 gm (weekly once)
Bear • Honey bee- weekly • Milk 500 ml • Radish 100 gm • Bread slices 2 no • Carrot 200 gm	**Jungle cat** • Chopped beef 250 gm • Bread slice 2 no. • Milk 100 ml **Otter** • Fish 1.5 kg • Crab (weekly twice) 300 gm • Sardine 1.5 kg.
Primates	
Bonnet macaque • Rice 50 gm • Banana 3 no. • Orange/Mango no. (100-150 gm size) • Guava no • Groundnut (WOS) 15 gm • Bengal gram 15 gm	

Common Diet Chart for Birds	
Love Birds (Peach faced love bird, Fisher's love bird, Sky blue mask love bird, Love bird i.e. Split)) Apple (to all) 1 no • Banana (to all) 1 no • Thinai 10 gm • Paddy 5 gm • Mixed grain 10 gm • White Bengal gram 5 gm • Onion 10 gm • Greens 10 gm • Shell grit Q.S.	**Indian Peafowl** • Cabbage 25 gm • Mixed grains 25 gm • Paddy 25 gm • White Bengal gram 50 gm • Greens 100 gm • Garlic 10 gm • Groundnut (WOS) 50 gm • Shell grit 25 gm • Green gram 25 gm
Parakeet • Bread slice 1 no • Apple (L.S) 3 no • Sathukudi/Mango/Orange no • Guava 1 no • Groundnut (WOS) 20 gm • Chilly fruit 10 gm • Garlic 2 gm • Paddy 10 gm • Mixed grains 20 gm • Bengal gram 10 gm • Sunflower seed 10 gm • Green grain 10 gm	**Emu** • Chopped beef or rat 150 gm/2no/day • (each 100-150 gm size) • Chicks 10 gm (once in a week) • Green gram 250 gm • Wheat 250 gm (soaked) • Broiler finisher grumble diet 250 gm • Rice 250 gm (each 100-150 gm size) • Milk 100 ml (boiled) • Tomato 100 gm • Green banana 2 no
Owl • Chopped beef or rat 150 gm/2no/day • (each 100-150 gm size) • Chicks 10 gm(once in a week)	**Budgerigar** • Hinai 10 gm • Greens 10 gm • Shell grit 20 gm • Bengal gram 5 gm • Green gram 5 gm

Comprehensive Feeding Schedule for some Wild Animals under Captivity

Animal	Average Body Weight (kg)	Diet	Amount of Feed per Day (kg)	Supplements	Amount of Supplements	Feeding Frequency	Special Considerations
Lion	150-250	Meat (beef, horse meat), bones	8-10 kg meat, 1-2 kg bones	Mineral mix	20-30 g/day	5-6 times per week	Include fasting days, vitamin A and E supplementation recommended
Elephant	3000-5000	Hay, fresh grass, fruits (bananas, apples), vegetables	100-150 kg total	Mineral mix, Vitamin E, Selenium	50-100 g mineral mix, 1000 IU Vitamin E/day, 1-2 mg Selenium/day	2-3 times per day	Ensure adequate calcium and phosphorus, provide salt lick
Giraffe	700-1200	Leaves, hay, browse (acacia), fruits	30-40 kg total	Calcium, Vitamin D_3	50-75 g calcium, 3000-5000 IU Vitamin D3/day	2 times per day	High-fiber diet, avoid excess grains, vitamin D essential
Zebra	300-400	Hay, fresh grass, pelleted feed	10-15 kg hay, 1-2 kg pellets	Salt, Vitamin E	25-50 g salt, 1000-2000 IU Vitamin E/day	2-3 times per day	High-fiber diet essential, monitor hoof condition
Chimpanzee	30-60	Mixed fruits, vegetables, nuts, boiled eggs	2-3 kg total	Multivitamins, Vitamin C	100-200 mg Vitamin C/day, 10-20 g multivitamin/day	3-4 times per day	Provide variety, vitamin C essential for primates
Giant Panda	70-160	Bamboo (primary diet), fruits, high-fiber biscuits	12-15 kg bamboo, 1-2 kg fruits	Calcium, Vitamin D_3	30-50 g calcium, 1000-3000 IU Vitamin D3/day	Constant access to bamboo, fruits 1-2 times daily	Monitor for dental health, fresh bamboo daily

Animal	Average Body Weight (kg)	Diet	Amount of Feed per Day (kg)	Supplements	Amount of Supplements	Feeding Frequency	Special Considerations
Penguin	5-35	Fish (herring, capelin), vitamin supplements	0.5-1.5 kg fish	Vitamin B Complex, Iodine	5-10 mg Vitamin B complex/day, 0.1-0.2 mg Iodine/day	2-3 times per day	Feed in water, monitor for iodine deficiency
Kangaroo	20-85	Fresh grass, hay, pelleted kangaroo feed, vegetables	2-4 kg total	Calcium, Phosphorus, Vitamin E	20-30 g calcium, 10-20 g phosphorus, 500-1000 IU Vitamin E/day	2 times per day	Low-starch diet, monitor bone health
Grizzly Bear	180-360	Fish (salmon, trout), fruits, meat, nuts	10-15 kg total	Vitamin A, Vitamin E	10,000-20,000 IU Vitamin A/day, 1000-2000 IU Vitamin E/day	2-3 times per day	Seasonal diet variation, reduce intake during winter
Polar Bear	350-700	Fish, meat, fruits, vegetables	15-20 kg total	Vitamin D, Omega-3 fatty acids	2000-5000 IU Vitamin D/day, 1-2 g Omega-3/day	2-3 times per day	High-fat diet preferred, especially in colder climates
Rhinoceros	2000-3000	Hay, fresh grass, vegetables, browse	50-70 kg total	Calcium, Phosphorus, Vitamin E	50-100 g calcium, 25-50 g phosphorus, 1000-2000 IU Vitamin E/day	2 times per day	High-fiber diet, avoid overfeeding grains

CHAPTER 41

VISIT TO POULTRY FARMS

Poultry farming is an industry. The latest techniques of automatic feeding not only save the labour problem but is also providing round the clock uniform supply of feed to the birds. The feed ingredients received at the farm should be got tested from the reputed laboratory to formulate balanced rations.

Objectives:

1. To acquaint the students with various operations of the poultry farm.
2. To acquaint the students with feed processing techniques.

Procedure:

The teacher will arrange a visit to a poultry farm and the students will be writing a report of various types of operation and feed formulation being used by the poultry farmer.

Report:

ANNEXURE-I

Chemical Composition of Common Feed Ingredients (ICAR, 2013)

1. Roughages (Dry Forages and Fibrous Feedstuffs)

Feedstuff	DM (%)	CP (%)	CF (%)	EE (%)	Ash (%)	TDN (%)	Ca (%)	P (%)
Green Fodder								
- Berseem	18-20	18.5	28.0	2.2	8.9	55.6	1.75	0.38
- Napier Grass	20-25	9.2	30.2	2.1	9.5	54.7	0.38	0.32
- Hybrid Napier	20-25	11.5	29.5	2.3	10.0	55.0	0.40	0.35
- Sorghum Green	20-25	7.5	29.0	1.8	9.0	50.0	0.25	0.30
- Lucerne Green	20-25	18.0	25.0	2.5	10.5	62.0	1.30	0.30
- Alfalfa	20-25	20.0	26.0	2.0	9.0	60.0	1.20	0.30
- Clover	20-25	16.0	28.0	2.5	8.5	58.0	1.00	0.32
Dry Fodder								
- Wheat Straw	88-90	4.0	41.0	1.5	8.0	43.0	0.22	0.12
- Rice Straw	88-90	4.0	37.0	1.7	9.0	41.0	0.25	0.13
- Groundnut Haulms	88-90	9.5	30.0	2.0	12.0	54.0	1.10	0.25
- Oat Straw	88-90	7.0	34.0	2.5	9.0	49.0	0.35	0.20
- Barley Straw	88-90	6.5	36.0	2.0	8.5	47.0	0.30	0.22
- Pea Straw	88-90	9.0	32.0	2.2	8.5	52.0	0.40	0.25
- Chickpea Straw	88-90	8.5	33.0	2.0	9.0	51.0	0.35	0.28
Legume Hay								
- Lucerne Hay	85-88	17.5	26.0	2.2	9.5	55.0	1.40	0.30
- Cowpea Hay	85-88	15.0	27.0	2.0	10.0	55.0	1.30	0.28
- Gram Straw	85-88	8.0	30.0	2.5	11.0	52.0	1.20	0.30
- Lentil Straw	85-88	7.5	29.0	2.2	10.5	50.0	1.10	0.28
- Pea Hay	85-88	16.0	28.0	2.0	9.0	54.0	1.00	0.32

2. Concentrates (Energy and Protein Sources)

Feedstuff	DM (%)	CP (%)	CF (%)	EE (%)	Ash (%)	TDN (%)	Ca (%)	P (%)
Cereal Grains								
- Maize	86-88	9.0	2.5	4.0	1.8	88.0	0.03	0.35
- Barley	88-90	12.0	6.0	2.5	2.5	77.0	0.06	0.40
- Sorghum (Jowar)	88-90	11.0	2.8	2.5	2.0	80.0	0.05	0.32
- Oats	88-90	11.5	12.0	4.5	3.0	72.0	0.15	0.38
- Wheat	88-90	13.0	2.0	2.5	2.0	87.0	0.05	0.45
- Rye	88-90	11.5	6.0	3.0	2.0	76.0	0.10	0.38
Cakes & Meals								
- Soybean Meal	88-90	44.0	6.0	1.8	6.0	77.0	0.30	0.70
- Groundnut Cake	88-90	48.0	10.0	8.0	6.5	72.0	0.20	0.50
- Mustard Cake	88-90	37.0	11.0	7.5	8.0	71.0	0.70	0.85
- Cottonseed Cake	88-90	23.0	22.0	6.5	6.0	71.0	0.20	0.65
- Linseed Cake	88-90	31.0	9.0	7.5	7.0	74.0	0.25	0.55
- Sesame Cake	88-90	31.0	15.0	7.0	8.5	70.0	0.40	0.55
- Sunflower Meal	88-90	33.0	30.0	8.0	7.5	70.0	0.20	0.55
Agro-Industrial By-products								
- Wheat Bran	88-90	14.0	11.0	3.5	6.5	68.0	0.15	1.20
- Rice Bran (Full Fat)	88-90	14.0	13.0	13.0	9.0	72.0	0.08	1.30
- Rice Polish	88-90	13.0	11.0	12.0	10.0	75.0	0.10	1.10
- Molasses	75-80	4.0	1.0	6.0	8.0	60.0	0.60	0.15
- Distillers Dried Grains	90-92	27.0	9.0	8.5	6.5	73.0	0.20	0.75
- Beet Pulp	85-90	8.0	11.0	0.5	5.5	60.0	0.40	0.25
- Apple Pomace	85-90	5.0	14.0	0.5	8.0	55.0	0.30	0.20
Oilseeds By-products								
- Coconut Cake	88-90	21.0	15.0	12.0	7.5	68.0	0.20	0.40
- Palm Kernel Cake	88-90	16.0	25.0	8.5	8.0	62.0	0.30	0.60
- Sunflower Cake	88-90	29.0	25.0	7.5	7.0	70.0	0.25	0.60

ANNEXURE-II

Nutrient content of few important available feed ingredients

Ingredient	ME (Kcal/ kg)	Crude protein (%)	Crude fiber (%)	Ether extract (%)	Lysine (%)	Methi-onine (%)	Calcium (%)	Phos-phorus (%)
Cereals								
Maize	3309	9.2	3.0	3.8	0.18	0.15	0.25	0.40
Bajra	2950	12.7	4.0	3.0	0.43	0.20	0.13	0.72
Wheat	3045	10.3	3.0	2.5	0.47	0.21	0.19	1.12
Rice kani	2345	7.9	8.0	2.5	0.33	0.17	0.11	0.48
Barely	2618	12.0			0.60	0.20	0.29	0.64
Cereal by product								
Rice polish	2837	12.7	12.0	16.0	0.55	0.24	0.27	1.37
De-oiled rice polish	2135	14.1	14.0	1.0	0.44	0.31	0.37	1.80
Wheat bran	1286	17.2	10.9	3.0	0.53	0.09	0.19	1.12
Vegetable protein								
Soybean meal	2300	45.0	6.0	1.0	2.57	0.76	0.36	0.90
Full fat soybean	3275	36.5	5.5	18.0	2.35	0.53	0.25	0.58
Groundnut cake	2700	38.0	9.0	8.0	1.6	0.42	0.16	0.16
Groundnut ext.	2128	40-42	11.2	1.0	1.47	0.68	0.31	0.67
Sunflower cake	1600	28.0	26.0	1.0	1.00	0.60	0.37	0.30
Mustard cake	2373	35.1	9.6	8.0	1.14	0.42	0.89	1.78
De-oiled mustered cake	1672	31.7	11.6	2.3	1.37	0.51	0.58	0.82
Cotton seed meal	1556	25.9	21.3	0.9	1.07	0.41	0.52	0.86

Ingredient	ME (Kcal/ kg)	Crude protein (%)	Crude fiber (%)	Ether extract (%)	Lysine (%)	Methi-onine (%)	Calcium (%)	Phos-phorus (%)
Guar meal	1821	37.1	10.5	6.7	2.58	0.39	0.41	0.50
Maize gluten meal	3300	60.0	2.0	2.5	1.02	1.28	0.22	0.19
Brewer's dried grains	2732	28.2			0.98	0.43	0.29	0.54
Animal proteins								
Fish meal	1834	43.1	2.5	6.0	2.5	1.08	7.16	2.5
Meat meal	2300	56.2	2.6	6.8	4.00	0.84	2.68	2.06
Meat cum bone meal	2100	48.0	-	8.0	3.72	0.75	11.25	5.39
Mineral source								
DCP	-	-	-	-	-	-	23	18
Limestone	-	-	-	-	-	-	35.0	-
Oyster shell	-	-	-	-	-	-	38.0	-

ANNEXURE-III

Calcium and phosphorus content of common mineral supplements

Source	Ca (%)	P (%)
Tri-calcium phosphate (TCP)	34.20	17.70
Di-calcium phosphate (DCP)	24.80	18.80
Mono-calcium phosphate (MCP)	16.90	21.50
Defluorinated rock phosphate	33.00	18.00
Calcium carbonate	40.00	-
Oyster shell grit	38.00	-
Calcium chloride anhydrate	36.00	-
Calcium chloride dihydrate	27.00	-
Calcium sulphate anhydrate	29.00	-
Calcium sulphate dihydrate	23.00	-
Dolomite limestone	-	22.00
Bone meal (steamed)	27.00	12.00
Mono-sodium phosphate	-	22.50
Di-sodium phosphate	-	08.70

(NRC-2012; Sauvant et al., 2004)

Calcium and phosphorus concentration in animal feed ingredients

Source	Ca (%)	P (%)
Blood meal	0.05	0.21
Blood plasma	0.13	1.28
Fish Meal	4.28	2.93
Meat meal	6.37	3.16
Meat and bone meal	10.94	5.26
Milk skimmed, dried	1.27	1.06
Poultry by-product meal	4.54	2.51
Whey powder, dried	0.62	0.69
Whey permeate, dried	0.27	0.34

(National Research Council, 2012)

ANNEXURE IV

Classification of Roughage and Concentrates

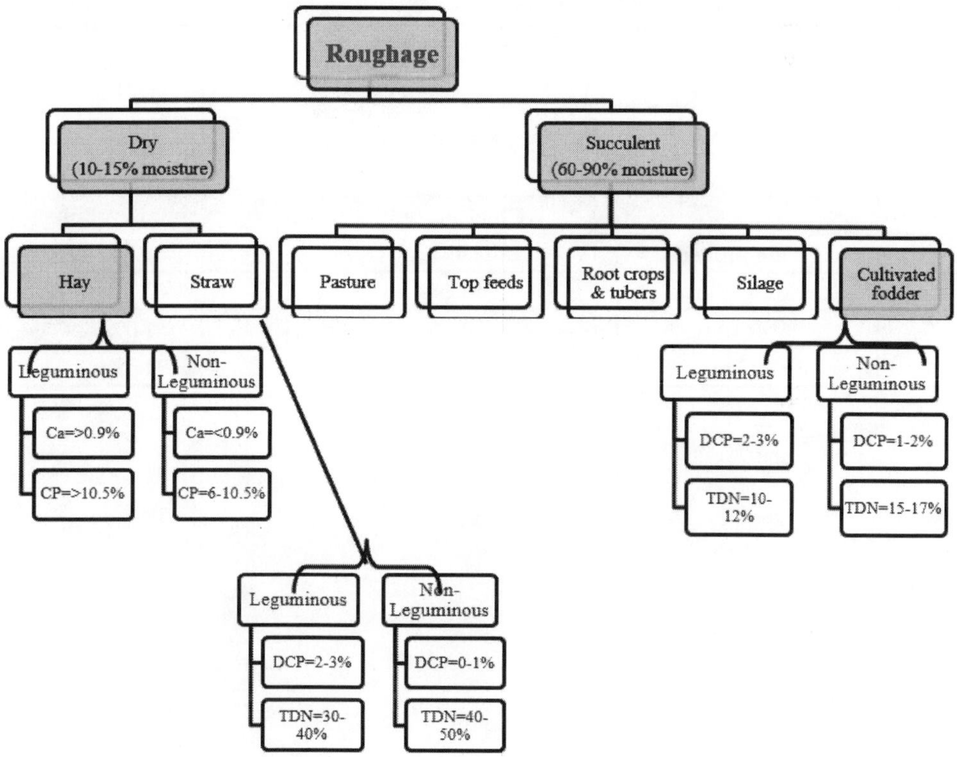

Other dry roughages:

- Corn cobs
- Cottonseed hulls
- Sugarcane bagasse

Miscellenous succulent roughages: Garbage, cull fruits, vegetables, waste from food processing plants, etc.

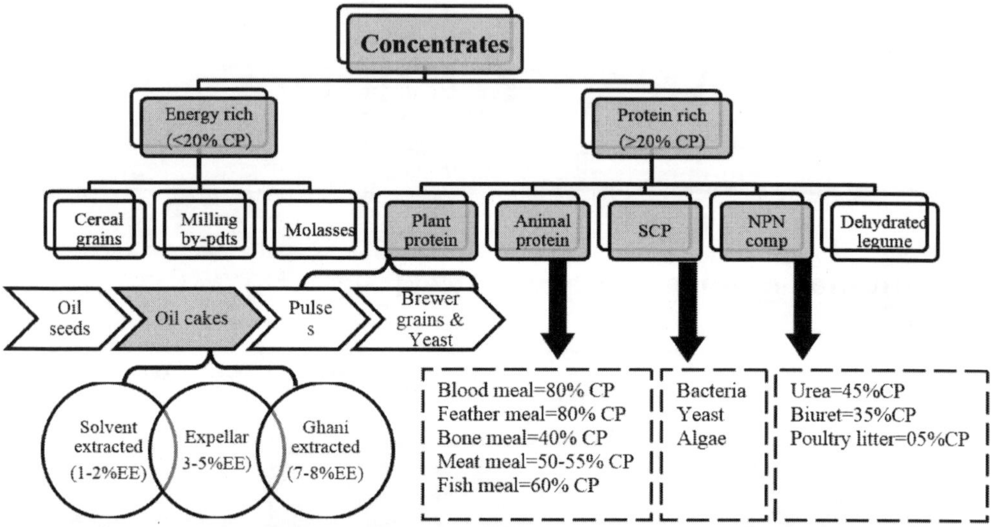

S.No	Character	Plant Protein	Animal Protein
1.	Crude protein	<47%	>47%
2.	Crude fiber	>2.5%	<2.5%
3.	Calcium	<1.0%	>1.0%
4.	Phosphorus	<1.5%	>1.5%

ANNEXURE V

Maximum prescribed limit for harmful substances in chicken feeds (BIS, 2024)

S. No.	Substance	Max permissible limit
i)	Arsenic	2 ppm
ii)	Fluorine	30 ppm
iii)	Lead	5 ppm
iv)	Mercury	Absent
v)	Nitrite	15 ppm
vi)	Aflatoxin B1	20 ppb
ix)	Free gossypol	20 ppm
x)	Hydrocyanic acid	10 ppm
xi)	BHC	20 ppb
xii)	DDT	55 ppb
xiii)	Endosulphan	10 ppb
xiv)	Aldrin	1ppb

Composition of Commonly Used Feed Ingredient (BIS, 2024)

S. No.	Ingredients	ME, Kcal/ kg	Proximate Specification (percent)					Amino Acids Specification (percent)										
			Crude Protein	Crude Fat	Crude Fibre	Total Ash	Sand Silica	Lysine		Methionine		Methionine+ Cystine		Tryptophan		Threonine		Linoleic Acid
								To-tal	Di-gest	To-tal	Di-gest	To-tal	Digest	To-tal	Di-gest	To-tal	Di-gest	
(1)	(2)	(3)	(4)	(5)	(6)	(7)	(8)	(9)	(10)	(11)	(12)	(13)	(14)	(15)	(16)	(17)	(18)	(19)
	Energy Ingredients																	
i)	Maize	3300	9	4	2	1.5	1	0.27	0.22	0.17	0.16	0.37	0.34	0.08	0.06	0.31	0.25	1
ii)	Jowar	3000	10	3	4	3	1	0.22	0.16	0.17	0.15	0.38	0.28	1.0	0.07	0.31	0.24	1.1
iii)	Bajra	2640	12.7	4.9	2.2	2	1	0.42	NA	0.24	NA	0.38	NA	0.18	NA	0.44	NA	NA
iv)	Rice (broken)	2600	7.9	1.7	1.4	4	2	0.26	0.21	0.2	0.17	0.37	0.31	0.09	NA	0.26	0.21	0.6
v)	Wheat	3100	14	2.6	2.5	2	1	0.38	0.31	0.21	0.18	0.5	0.44	0.15	NA	0.38	0.32	2
vi)	Barley grain	2640	11.5	1.9	5	2.5	1	0.44	0.34	0.18	0.15	0.42	0.34	0.12	NA	0.53	0.45	1
vii)	Rice polish	2700	12.7	14	5	8	2.5	0.54	0.4	0.24	0.18	0.5	0.36	0.14	NA	0.45	0.31	4.4
viii)	Ragi	2950	12.6	1.85	2.8	1.45	0.5	0.36	0.24	0.15	0.2	0.36	0.26	-	-	0.32	0.3	-
ix)	Tapioca flour	3300	2	1.7	12	8.5	-	006	-	0.006	-	0.01	-	-	-	-	-	-
x)	Vegetable fat	8800	-	99.4	9	-	-	-	-	-	-	-	-	-	-	-	-	31.0
	Protein Sources																	
xii)	Soy extract	2500	48	1	3.5	5	2	2.98	2.68	0.69	0.63	1.41	1.24	0.61	0.57	1.89	1.68	0.4
xii)	Soy meal	2250	44	0.8	6.5	6	2.5	2.75	2.48	0.64	0.59	1.31	1.15	0.57	NA	1.76	1.57	0.4
xiii)	Groundnut extract	2690	48	1.5	6.8/	7.2	2.5	1.77	NA	0.42	NA	1.15	NA	0.5	NA	1.16	-	0.19
xiv)	Groundnut meal	2400	44	1	10	8	2.5	1.39	1.07	0.42	0.37	1.1	0.9	0.41	NA	1.12	0.94	0.19
xv)	Rapeseed extract	1900	36	1.7	11.5	7	2	2.02	1.61	0.73	0.65	1.64	1.33	0.47	0	1.58	1.26	0

S. No.	Ingredients	ME, Kcal/ kg	Proximate Specification (percent)					Amino Acids Specification (percent)										
			Crude Protein	Crude Fat	Crude Fibre	Total Ash	Sand Silica	Lysine		Methionine		Methionine+ Cystine		Tryptophan		Threonine		Linoleic Acid
								To-tal	Di-gest	To-tal	Di-gest	To-tal	Digest	To-tal	Di-gest	To-tal	Di-gest	
xvi)	Sunflower extract	1540	28	1	24	7.7	2	1.06	0.86	0.67	0.62	1.15	0.99	0.34	0	1.05	0.89	0.5
xvii)	DDGS(rice)	2883	46	4.5	4.9	10.2	4.28	1.34	0.94	1.13	0.66	2.01	1.09	0.55	0.36	1.66	1.03	-
xviii)	DDGS(corn)	2617	38	10.7	6.0	4.3	2.0	1.11	0.72	0.87	0.75	1.57	1.33	0.41	0.33	1.35	0.97	-
xix)	Sesame extract	2200	44	0.5	6.1	11.5	2	1.01	0.83	1.16	0.98	1.97	1.6	0.54	0	1.44	1.05	1.9
xx)	Full fat soy	3300	38	18	5	4.6	2	2.37	1.9	0.51	0.4	1.16	0.83	0.69	0.45	1.57	1.21	7.7
xxi)	Maize gluten 42	3150	42	2	4	3	1.5	0.8	0.63	1.0	0.85	1.6	1.33	0.2	0.17	1.1	1.06	-
xxii)	Maize gluten 60	3650	60	2	2.5	1.3	0.5	1.1	0.85	1.5	1.3	2.3	2.0	0.3	0.25	1.6	1.32	-
xxiii)	Fish meal	2180	45	7	1	22	5	2.5	2.8	0.84	0.7	1.15	0.86	0.7	-	1.4	1.2	-
xxiv)	Meat and bone meal	1848	45	8.6	2.1	38	-	2.48	2.12	0.65	0.59	1.16	0.94	0.29	0.23	1.6	1.36	0.3
xxv)	Rice bran doc	1800	16	0.5	14	12.5	5	0.66	0.48	0.31	0.23	0.64	0.44	0.17	-	0.54	0.37	-
xxvi)	Wheat bran	1400	14.5	3	11	7	2	0.63	2.12	0.23	0.19	0.56	0.44	0.28	0.21	0.52	0.38	1.7
xxvii)	Salseed extract	2500	9	1	3	-	1.5	0.5	0.38	0.32	0.24	0.52	0.4	0.1	0.07	0.3	0.23	0.3
xxviii)	Molasses	2000	3	-	-	9.5	-	-	-	-	-	-	-	-	-	-	-	-

Notes: 1 NA—Analysis not available.

2 All the above values are on DM basis.

Composition of Commonly Used Feed Ingredients: Natural Vitamins (mg/kg) as per BIS, 2024

Sl No.	Raw Material	Ribo-flavin	Nia-cin	Pantothenic Acid	Thia-mine	Cho-line	Bio-tin	Folic Acid	Pyri-doxine	Caro-tene	Xantho-phyll	Alfa Tocopherol	Vitamin A IU/kg
(1)	(2)	(3)	(4)	(5)	(6)	(7)	(8)	(9)	(10)	(11)	(12)	(13)	(14)
	Energy Ingredients												
i)	Maize	1.10	19.80	3.96	3.3	506	0.07	0.29	4.62	1.98	154	11.0	3300
ii)	Jowar	1.20	42.70	11.00	3.9	678	0.18	0.02	NA	NA	NA	NA	0
iii)	Bajra	1.60	52.60	7.40	6.6	789	NA	NA	NA	NA	NA	NA	NA
iv)	Rice (broken)	0.88	28.60	7.92	2.64	792	0.07	0.18	4.40	0	0	8.80	0
v)	Wheat	1.10	48.40	9.02	4.4	1078	0.11	0.42	2.20	0	0	8.80	0
vi)	Barley grain	1.54	52.80	7.92	4.84	1100	0.04	0.66	2.86	0	0	7.04	0
vii)	Rice polish	1.76	407.00	46.20	19.8	1232	0.62	NA	13.20	NA	NA	90.20	NA
viii)	Ragi	1.50	15.60	9.20	4.4	0	0.06	0.06	0.00	0	0	0	0
ix)	Tapioca flour	0	0	0	0	0	0	0	0	0	0	0	0
x)	Vegetable fat	0	0	0	0	0	0	0	0	0	0	0	0
	Protein Sources												
xi)	Soybean meal	2.86	26.40	15.40	4.62	2640	0.33	0.66	5.94	-	-	1.96	-
xii)	Groundnut cake	7.70	165.00	46.20	5.94	1892	0.37	0.40	8.8	-	-	2.86	300
xiii)	Rapeseed meal	3.52	158.40	9.46	5.06	6699	0.88	2.20	7.04	-	-	-	-
xiv)	Sunflower cake	2.2	198.00	28.60	2.86	2860	0.66	1.32	11.00	-	-	11	7
xv)	DDGS (rice)	NA	NA	NA	NA	NA	NA	NA	NA	NA	NA	NA	NA

Sl No.	Raw Material	Ribo-flavin	Nia-cin	Pantothenic Acid	Thia-mine	Cho-line	Bio-tin	Folic Acid	Pyri-doxine	Caro-tene	Xantho-phyll	Alfa Tocopherol	Vitamin A IU/kg
xvi)	DDGS (corn)	9.0	79.9	11.4	3.5	3400	0.3	0.88	-	-	-	40	2700
xvii)	Sasame/til cake	3.30	28.60	5.50	2.64	1496	0.33	0.00	12.54	-	-	-	-
xviii)	Fully fat soy	2.64	22.00	11.00	4.62	2860	0.29	4.18	10.78	-	-	9.24	0
xix)	Maizegluten 42	1.50	54.50	9.60	0.22	330	0.15	0.22	3.6	16.2	-	-	25
xx)	Maizegluten 60	2.20	81.00	2.90	0.28	2200	0.22	0.23	2.8	44	-	-	60
xxi)	Fishmeal	6.60	66.00	8.80	1.32	3060	0.09	0.20	0	-	-	-	-
xxii)	Meat and bone meal (50 percent)	1.98	35.20	4.62	0.44	1320	0.04	0.22	-	-	-	-	-
xxiii)	Ricebrandoc	0.88	28.60	7.92	2.64	792	0.07	0.18	4.4	-	-	8.8	-
xiv)	Wheatbran	3.08	184.80	28.60	6.60	1870	0.22	1.9	6.82	0	0	13.2	0
xxv)	Salseedex-tract	NA	NA	NA	NA	NA	NA	NA	NA	NA	NA	NA	NA
xxvi)	Molasses	2.86	33	37.4	0.88	748	0.66	0.09	4.18	NA	NA	7.04	NA

Feed Raw Material Specification—Macro and Micro Minerals (BIS, 2024)

S. No.	Raw Materials	Macro Minerals (percent)								Micro Minerals (mg/kg)				
		Calcium	Total Phosphate	Avail Phosphate	Sodium	Chloride	Potassium	Magnesium	Sulphur	Copper	Iron	Manganese	Zinc	Selenium
(1)	(2)	(3)	(4)	(5)	(6)	(7)	(8)	(9)	(10)	(11)	(12)	(13)	(14)	(15)
	Energy Ingredients													
i)	Maize	0.01	0.25	0.08	0.01	0.04	0.42	0.11	1100	2	40	7	10	0.04
ii)	Jowar	0.04	0.40	0.13	0.01	0.1	0.4	0.13	900	10	40	10	10	0
iii)	Bajara	0.13	0.72	0.36	0.04	0	0.43	0.16	1300	21.6	40	29	13.9	0
iv)	Rice (broken)	0.11	0.48	0.24	0.04	0	0.34	0.14	50	3.3	40	17.6	1.8	0
v)	Wheat	0.05	0.28	0.14	0.02	0.05	0.35	0.15	1400	5	50	38	30	0.065
vi)	Barley grain	0.08	0.42	0.21	0.03	0.18	0.56	0.12	1500	10	70	16	15	0.35
vii)	Rice polish	0.27	1.37	0.14	0.06	0.07	1.7	0.95	1800	13	190	130	30	0
viii)	Ragi	0.08	0.30	0.00	0.00	0.00	0.00	0.00	0.00	0.00	0.00	0.00	0.00	0.00
ix)	Tapioca flour	0.58	0.12	0.00	0.00	0.00	0.00	0.00	0.00	0.00	0.00	0.00	0.00	0.00
x)	Vegetable fat	0.00	0.00	0.00	0.00	0.00	0.00	0.00	0.00	0.00	0.00	0.00	0.00	0.00
	Protein Sources													
xi)	Soybean meal	0.40	0.60	0.14	0.02	0.05	2.00	0.27	4300	21	100	27	40	0.01
xii)	Groundnut cake	0.10	0.60	0.20	0.07	0.00	1.15	0.33	2800	0	0	26	21.7	0

S. No.	Raw Materials	Macro Minerals (percent)								Micro Minerals (mg/kg)				
		Calcium	Total Phosphate	Avail Phosphate	Sodium	Chloride	Potassium	Magnesium	Sulphur	Copper	Iron	Manganese	Zinc	Selenium
xiii)	Rapeseed meal	0.60	1.10	0.25	0.03	0.10	1.20	0.64	2500	6	160	53	71	0.98
xiv)	Sunflower cake	0.35	0.90	0.30	0.01	0.05	1.00	0.62	3800	35	140	34	100	1.5
xv)	Sasame/ Til cake	2.00	1.30	0.24	0.04	0.00	1.39	0.00	0	0	0	48	0	0
xvi)	DDGS (rice)	0.73	0.77	0.52	NA	NA	NA	NA	NA	NA	NA	NA	67	NA
xvii)	DDGS (corn)	0.29	0.68	-	0.2	0.17	1.0	0.42	0.3	50	300	30	85	0.35
xviii)	Fully fat soy	0.25	0.59	0.20	0.04	0.08	1.70	0.21	3000	15	75	30	40	0.1
xix)	Maize gluten 42	0.16	0.40	-	0.10	-	0.03	0.06	600	28.2	400	73	-	1
xx)	Maize gluten 60	0.02	0.70	-	0.03	-	0.45	0.15	600	22	167	4.4	41	1
xx)	Fish meal	7.16	1.67	1.67	0.19	0.60	-	-	-	21.7	320	38.9	-	0
xxi)	Meat and bone meal (45percent)	10.00	5.10	5.10	0.60	0.00	1.30	1.00	2600	0	500	10.1	0	0.25

S. No.	Raw Materials	Macro Minerals (percent)								Micro Minerals (mg/kg)				
		Calcium	Total Phosphate	Avail Phosphate	Sodium	Chloride	Potassium	Magnesium	Sulphur	Copper	Iron	Manganese	Zinc	Selenium
	Others													
xxii)	Rice bran doc	0.37	1.80	-	0.06	0.07	1.70	0.95	1800	13	190	130	30	0
xxiii)	Wheat bran	0.08	1.15	0.40	0.02	0.06	1.23	0.50	2200	14	170	100	9.5	1
xxiv)	Sal seed extract	0.24	0.16	0.00	0.00	0.00	0.00	0.00	0.00	0.00	0.00	0.00	0.00	0.00
xxv)	Molasses	0.80	0.08	0.04	0.16	2.31	2.8	0.35	3500	50	180	40	20	0
	Minerals													
xxvi)	Lime stone powder	38.00	0.00	0.00	0.05	0.03	0.10	2.10	0	0	3.5	0	0	0
xxvii)	Dicalcium phosphate	23.00	18.00	18.00	0.06	0.01	0.10	0.60	11 100	70	500	120	380	0
xxviii)	Sodium chloride	0.30	0.00	0.00	39.00	60.00	0.00	0.01	2000	0	50	0	0	0
xxix)	Soda bicarbonate	-	0.00	0.00	27.00	0.00	0.00	0.00	0	0	0	0	0	0
xxx)	Shell grit	38.00	0.10	0.10	0.20	0.01	0.10	0.30	0	0	2.9	100	0	0
xxxi)	Bone meal	29.80	12.50	12.50	0.04	0.00	0.20	0.30	24 000	0	5	0	100	0

Note—NA- Analysis not available.

INDEX

V

W